Destructive Interference
Understanding the brain's telepathic potential

Mark Fox BSc (Hons) First Class

Paperback edition first published in the United Kingdom in 2014 by aSys Publishing

eBook edition first published in the United Kingdom in 2014 by aSys Publishing

Second Edition published in the United Kingdom in 2021 by aSys Publishing

Hardback edition first published in the United Kingdom in 2021 by aSys Publishing

A CIP catalogue record for this book is available from the British Library.

ISBN: 978-1-913438-45-6

aSys Publishing 2021

http://www.asys-publishing.co.uk

There are intangibles which can't be taught, notwithstanding that teaching can corrupt the most intangible of things. For the record, Destructive and Constructive Interference have intangible origins, having been commenced at the behest of one's heteronnubial co-author, who's brought both style and substance to their telling. Their balanced-minded presence proving – not simply that it's possible to experience another's heartbeat – but also that a renaissance woman's legendary perspective can be psychically arrived at and consolidated evocatively in writing.

Contents

Preface

First published in 2014, *Destructive Interference: Understanding the brain's telepathic potential* focusses on the intrigues surrounding the establishment of today's scientific orthodoxies. By openly acknowledging growing incommensurability within those orthodox 'sciences', this groundbreaking book was able to establish such concepts as *Coulomb's solution*, *epistemological jurisprudence* and the *dark cycle*. Accordingly, *remote manipulation*, *mental telepathy* and the *superimposition of persons* were all effortlessly explained, and those nagging conceptual inconsistencies systematically resolved. As for the neologism '*heteronnubial*', that denotes the cyclical convergence of both a man and woman's souls when married together in the above manner. Thus, this seminal evaluation, and its timely sequel – entitled *Constructive Interference: Developing the brain's telepathic potential* (aSys Publishing, 2019) – are the product of two convergent minds.

Additionally, in deference to the concept of *vicarious liability*, whereby criminal responsibility extends well beyond those immediately accountable, this *second edition* also includes a feminist studies 'blog pitch', entitled *Child, Early and Forced Marriage: A Crime of Omission?* Which serves as an emphatic warning to anyone occupying a commanding position not to be complacent about human rights, especially in an age of burgeoning extrasensory intrusions. In fact, all who appear 'cavalier' – including the English Crown – ought to be treated with suspicion, lest their application prove criminally unsound. For, as the book presciently asserts: "*Managing power wisely is as much about the orderly transfer of power as it is about its prudent application*" – such that anyone who's failed to grasp that the old order is epistemically deficient, simply hasn't woken-up to the gravity of the powers presently at humankind's disposal.

PART I

Science

Chapters 1-3

Ch1:   The philosophy of science

•   Paradigm postponed

1965 was a particularly intense year as regards the philosophy of science, and the two great names who helped to make that year so stimulating were Sir Karl Popper (1902-94) and Thomas Kuhn (1922-96). In that year, in London, the fourth International Colloquium on the Philosophy of Science devoted much of its time to critically appraising Kuhn's recent book, called The Structure of Scientific Revolutions (1962). Chairing that conference was the recently knighted philosopher, Karl Popper. At stake was our whole perception of what science does, how science advances, and, perhaps most significant of all, what science ought to be doing.

Many years earlier, in 1934, Popper had published his landmark work – entitled The Logic of Scientific Discovery (*Die Logik der Forschung*) – which stressed the importance of falsifiability in science. According to Popper the refutation, repudiation or **falsification** of a statement or hypothesis carried more intrinsic weight, scientifically-speaking, than the much flimsier concept we call **induction**, whereby we arrive at a conclusion on the basis of simple observation alone. The problem with simple observation, and hence induction, is that truth and reality are, by their very nature, often deceptive; and the role of the scientist must be to see through these illusions, not merely amplify the same.

Thus falsification, or **empirical refutation**, were, according to Popper, the key to scientific advance. By way of example, he proffered the theory that "every day the sun will rise". Clearly, if the sun ever fails to rise then the theory is falsified. Otherwise, the sun continues to rise and we're left with not so much the truth as a **tentative answer**. In Popper's view science gifts us a whole series of tentative answers, in place of what we might term the truth. This book seeks to convey the truth in a manner which is conscious of the illusory nature of reality, the tentative nature of knowledge, and the ever-growing demands on science in an age which transcends mere discovery.

The Popperian vision, presented above, has been taken as exemplifying the relatively slow, cumulative, one-directional path, that so many people believe the history of science has taken. This notion – that scientific advance is smooth and continuous, rather than convulsive or episodic – began to come under attack in the mid-1960s. Chief among the critics at that time was Thomas Kuhn,

2

whose book conceived of science as a whole series of paradigms. Kuhn saw the **paradigm** as a conceptual framework within which scientific research is carried out. The prevailing paradigm, at any given time, embodying what Kuhn would describe as **normal science**. Normal science might very well be Popperian in nature, i.e. limited in scale, incremental, one-directional, etc; but every so often the overarching conceptual framework, within which such science takes place, changes – sometimes dramatically. This sudden revolutionary repositioning of science was afforded the term **paradigm shift**. A paradigm shift is likely to be felt way beyond the confines of science itself, with practical implications for society, culture and even psychology. As a potential witness to such events, the reader may wish to familiarize themselves with the arguments contained within this text, as they are likely to be heavily conditioned by the paradigm which is ending.

This current paradigm is the exemplification of normal science, and as such is a fair reflection of what most scientists understand as proper science. Most scientists, after all, work well-within the parameters of the prevailing scientific zeitgeist, or spirit of the age. This climate, however, is a reflection of many factors, not all of them scientific. Political and military considerations have resulted, for example, in certain assumptions being actively reinforced, rather than questioned – thereby postponing the paradigm shift which would otherwise occur. This has happened over the course of the previous century, and we are now living in its shadow. A cruel and unusual shadow, cast by normal science's predilection for artifice.

There are, however, strict limits as regards postponing such a conceptual shift, due to growing **incommensurability** within normal science. Try as one might, discrepancies between prevailing theory and observation *will* force the issue into the open. Therefore, choosing precisely the right moment to embrace revolutionary change is key. And, capitalizing on such a moment is as much an act of political and military necessity as, for example, staying silent. This book argues that some within science have reinforced certain assumptions about the brain in order to keep the brain's secrets; trapping many, both inside and outside science, in an outdated conceptual framework. A conceptual framework in which people appear ever more technological, but by no means scientific.

3

• Science versus technology

Before continuing it's important to distinguish between science and technology. As the writer Cyril Aydon rightfully points out, science and technology are, respectively, mankind's way of explaining and exploiting natural forces.[1] In fact, technology often appears to lift scientific awareness above the realms of the merely tentative, affording such knowledge the status of unequivocal truth. In reality, the greatly simplified conceptual models which explain such technology remain tentative answers. Thus, normal science's heavy reliance on these rather abstract conceptual models is both a strength and a weakness.

The continuance of normal science and its associated models is no guarantee that a new paradigm isn't taking shape surreptitiously. By simply reinforcing commonly held beliefs about the brain, science – or, should I say, a faction within science – has been able to secretly develop technologies which are capable of **remote manipulation**, **mental telepathy** and the **superimposition of persons**. All of which harness some of the most fundamental chemical processes in nature – and, in ways that no simple process of induction would ever have suggested. Taken together, these three phenomena are collectively termed **telepathically-induced effects**.

Such effects are the result of **radiofrequency technologies**; which, by their very nature, provide for cognitive transference, remote manipulation of physiological processes, and the conscious control of third parties (using only the power of one's mind). Powerful though these technologies are they mustn't blind us to the **scientific method**, or the prudent pursuit of knowledge. Moreover, straightforward induction is never more wanting than when faced with the clandestine application of such powers.

The scientific method can be defined as follows: clinical research leads to the identification of a problem to be solved. A suggested solution follows, called the working **hypothesis**. This hypothesis is worded in such a way that it can be tested by experiment, ideally predicting the result, e.g. "all swans are white" (falsified by the sight of a black swan). If the prediction proves accurate it can be refined into a theory (which, at best, would be one of science's tentative answers). If repudiated, another working hypothesis is required, along with further experimentation. When the problem is finally resolved the work is written-up and published.

4

The foremost reason for adopting a falsifiable hypothesis which is consistent with one's expectations is that experimentation can be very expensive, and it is far better to arrive at a truly ground-breaking new theory, than almost certain refutation. This isn't unscientific, as the methodologies remain consistent, i.e. it's all hypothetico-deductive research. Technology, whatever its description, merely facilitates such research, it's not a substitute for the same. The danger we now face is that we've ended up under the influence of various devices, whose broad application pre-dates orthodox scientific research.

· **Reality as an illusion**

Popper's disquiet, as regards inductive forms of reasoning, lay with the fact that **nature** is often highly deceptive (in this context, the terms nature and reality are interchangeable). Not only are we more likely to be deceived by inductive forms of reasoning, the conclusions are more likely to be viewed as definitive. Induction, therefore, encourages us to make too many assumptions about nature, and provides for too many false conclusions. If one adopts the convention that any inquiry which is not inductive is deductive, and vice versa, then Popper's suggested methodology represents a valid form of **deduction**, by way of **hypothetico-deductive research**.[2] The question then becomes, is the present conceptual framework, or paradigm, good enough to crack nature's deepest codes?

This question will resonate far more acutely if I remind you that **electromagnetic radiation** (light in other words) appears *prima facie* to exert no pressure; to exist only in the visible; to act on things in a mostly uniform manner; to travel infinitely fast; to not travel in waves; and be wholly unaffected by **gravity**. When, in reality, such radiation does exert a pressure; has wavelengths beyond the visible; is immensely varied in its effects; travels at a finite speed; in the manner of a wave; and, in the case of the **photon**, possesses just enough **mass**, when not at rest, to be affected by gravitational forces. Reality is fraught with such illusory phenomena, which is why science must be so exacting.

Given the true nature of reality, scientific deception, at its simplest, merely involves inviting people to assimilate tentative answers as unquestionable truths. This is one of the reasons why paradigm shifts come as such a shock. As the prevailing paradigm ages, more and more people are invited to treat the resulting corpus of knowledge as definitive, making them victims of their own inductive thought-processes when revolutionary change finally arrives. Such change occurs because a scientific elite, wrestling with growing incommensurability, reaches for a whole new paradigm. If that new paradigm has already been arrived at in secret, then the transition may be even more abrupt.

Now imagine that your *prima facie* assumptions about the human brain have been reinforced for political and military purposes. It's not difficult to disguise certain aspects of the brain which are already heavily disguised by nature. Whenever science confirms your preconceptions, you're inclined to be pro-science. However, can you be absolutely certain that your *prima facie* assumptions about the human brain aren't being reinforced for political or military purposes? This book suggests that such a deception *has* occurred, and that people are more than willing to have their preconceptions about **neurology** reinforced, whatever the underlying realities. In that sense, such people are complicit in their own deception.

With this in mind, let's consider the process of scientific discovery as regards electromagnetic radiation; a term which is more often than not abridged to simply **light** or **radiation**. If the visible portion of the electromagnetic spectrum is spoken of, it's referred to as **visible light**. Whilst reading the following historical account, imagine any one of these contenders for a 'new paradigm' being quietly and disingenuously erased from common knowledge, and replaced with a model based entirely on public preconceptions. Bearing in mind just how wrong those preconceptions are likely to have been.

- ### Light appears to exert no pressure

Building on the work of others, the Danish astronomer Ole Rømer (1644-1710) carefully observed the movements of Io, one of Jupiter's moons. By 1676 (the year in which Rømer made his observations) it was understood that the Earth orbits the sun quicker than Jupiter. Rømer soon realized that as the Earth's distance from Jupiter increased, Io disappeared behind Jupiter later than expected; and, that as the Earth got closer to Jupiter this small moon would disappear sooner than one would expect. Applying the incisive logic known as **Ockham's razor**, Rømer postulated that light has a finite speed, and that this accounted for the perceived **anomaly**. In fact, we know today that light reflected from Io's surface has to travel an extra 200 million miles through space, when the Earth and Jupiter are furthest apart, its precise speed being 299,792,458 metres per second.

Contemporaneous with Rømer's astronomical observations was Christiaan Huygens' research into the wave-like properties of light. Huygens (1629-95) was a Dutch physicist, who published his revolutionary Treatise on Light in 1690. Implicit in his wave theory of light is the suggestion that space is an **isotropic medium**. In other words, that space is uniform in all directions, thus providing for the propagation of light-waves in a regular and predictable manner. **Reflection, refraction, diffraction** and even **interference** effects could thus be explained. Later work by the English physicist Thomas Young (1773-1829), and French physicist Augustin-Jean Fresnel (1788-1827), confirmed Huygens' thinking.

Young shone a light through a screen containing two slits and produced interference patterns on a second screen placed behind the first (an effect which could only be produced if light were a wave).[3] As for Fresnel, he was able to show that these waves were **transverse waves**, just like ocean waves, i.e. the waves form perpendicular to their line of travel. All such waves have a **wavelength** (the distance between successive crests), an **amplitude** (the height of the crest above the rest position or flat-calm), and a **frequency** (the number of waves emitted by the source in just one second). Frequency is normally measured in **hertz** (Hz); for example, a BBC transmitter emitting 97,600,000 radio waves per second produces a frequency of 97.6 MHz (i.e. Radio 1).

The **Huygens-Fresnel principle**, as this wave-theory of light became known, clashed with an alternative explanation of light's behaviour proposed by Englishman Sir Isaac Newton (1642-1727). Newton perceived light as behaving in the manner of so many "corpuscles" (i.e. particles), and that visible light comprises various colours, which can be discerned when white light is split with a prism. Although seemingly incommensurable, these two camps (namely, wave theory and particle physics) were eventually reconciled in the 20th century, via the concept we now know as **wave-particle duality**. But, whichever camp you belonged to, it was clear that sunlight comprised a whole spectrum of colours.

Joseph Von Fraunhofer (1787-1826), a German physicist who founded an optical institute in Munich in 1807, discovered, on closer inspection, that the sun's spectrum contained both emitted colours *and* dark absorption bands or lines, called **Fraunhofer lines**. Fraunhofer is therefore credited with being the inventor of the spectroscope, a device which exploits the fact that these emitted colours correspond with the presence of certain **elements** in the light source; and any dark lines with the absence of certain elements. Fraunhofer's exquisite technology helped to found the whole new science of **spectroscopy**. Spectroscopy being, without doubt, *the* most significant revolution of the entire French Revolutionary and Napoleonic era.

It didn't take long for people to start probing the sun's chemical composition using this technique. However, understanding the sun's chemical composition is complicated by the fact that the sun possesses an atmosphere. Any light emitted from its surface must pass through these gaseous outer layers before it reaches an observer here on the Earth. In keeping with the conceptual model known as a **black body**, the sun's surface emits in all the visible wavelengths (it's surface temperature being around 5,800^0 C). However, in spite of its surface emitting in every visible wavelength, dark absorption lines would nonetheless appear in the visible spectrum produced here on Earth. Something in the sun's atmosphere had to be absorbing those missing wavelengths.

German physicist Gustav Kirchhoff (1824-87) deduced that if light passes through a gas the resulting spectrum displays absorption lines in the same places as the emission lines of the same gas, when incandescent. Put more simply, a gas is likely to absorb light emitted by the same gas, provided the latter is hot enough to emit and the former is cool enough to absorb. In the case of the sun, various elements could be detected in its relatively cooler atmosphere using this method, and it became possible to estimate the sun's temperature by studying this information. Thanks to Kirchhoff, we now know that the outer layers of the sun's atmosphere are sufficiently cool for selective absorption of radiation from its surface to occur.

Contemporaneously, in 1868, the English astronomer Norman Lockyer (1836-1920) detected the presence of an unknown element in the atmosphere of the sun, using simply its emission spectra (in this case a single yellow line). He called this mysterious element helium. A whole generation later, in 1895, the gas helium was successfully isolated in a laboratory on earth. Shortly after helium's initial discovery a revolutionary new way of looking at light emerged – one which would alter the entire course of human history.

In 1873 the Scottish physicist James Clerk Maxwell published his magnum opus, entitled Treatise on Electricity and Magnetism. In this monumental work, Maxwell showed that electricity, magnetism and light were all aspects of the same natural phenomenon, i.e.

electromagnetism. Electrical currents appeared to generate magnetic fields; but equally, magnetic fields appeared to produce electricity. The ability of magnetic fields to both strengthen and weaken one another is central to this book's broader arguments. Thus, by Victorian times, the electromagnetic spectrum, light, electricity, magnetism, and even brain function, were all starting to appear linked – brain waves having been discovered in 1875. Even the crudest of electrical currents appeared to be capable of producing electromagnetic waves; but precisely how these could be detected and harnessed remained an open-question. Then, in 1887, German physicist Heinrich Hertz (1857-94) became the very first person to transmit and receive long-wave radiation, otherwise known as radio waves. So much progress had been made by this time that light was seen simply as an aspect of electromagnetism, something which arose spontaneously due to the electrical properties of matter. As for **radiation pressure**, the pressure exerted on an object by light, that was demonstrated experimentally in 1900 by the Russian physicist Peter Lebedev (1866-1912). The unthinkable now appeared possible; in an age still heavily scarred by centuries old competition and imperial warring.

- **Telepathy doesn't appear to exist**

The **Russians**, Germans, French, English, Americans and Scots were all converging on the deeper mysteries of matter at a time when British elder statesman Joseph Chamberlain was warning "All Europe is armed to the teeth" and that "under such conditions the weak invite attack". It was as though the countries of Europe were hedging their bets, placing their faith in military prowess, should their science fail to compete with the competition. This was an age when theoretical physics was struggling to account for the advances in applied physics, and it was beginning to look as though almost anything was possible. And, without any question of a doubt, the country the English-speaking world feared the most was Germany.

The Congress of Vienna, which reconstructed Europe following Napoleon's downfall in 1815, created a loosely articulated confederation of Germanic states. States which finally came together as a coherent German nation in 1871, following the wars of unification. German nationalists were committed to forging a formidable German homeland – one which could compete with both Britain and France overseas, becoming a major economic and military power in the process. Any scientific or technological breakthroughs achieved by Germany at this time would undoubtedly have been used to pursue Germany's imperialist cause. Therefore, German ascendancy became all the more threatening when seen through the eyes of the English-speaking scientific elite.

It was within this climate that the world's foremost powers began to polarize into two distinct camps: specifically, the Allies and the Central Powers. The Allies comprised Britain, France and **Russia**; and the Central Powers comprised Germany, Austria-Hungary and Turkey. Between 1914 and 1918 these two immense power blocks became locked in total war, with Russia's support for Serbia providing the trigger. The assassination of the heir to the Austrian throne by a Serb nationalist, in June 1914, led to Austria-Hungary declaring war on Serbia with the full-backing of Germany. Accordingly Britain, France and Russia, together with all their respective allies, became mortal enemies of Germany, and those allied to Germany. Amid this cult of militarism and associated bloody conflict the whole philosophy of science began to be shaped not by openness, but by intrigue.

Germany hadn't anticipated total war, having gambled instead on incapacitating its chief rivals through a rapid invasion of France, followed by a secondary attack on Russia. In the event, the German advance ground to a halt along a line running from Switzerland to the North Sea (the so-called Western Front). Nonetheless, Germany went ahead with its planned attack on Russia, resulting in an Eastern Front. In addition to these two fronts the war was also fought in the **Middle East** (against Germany's ally Turkey), in Africa, and at sea. Today, the received wisdom is that if the Great War had been fought 20 years earlier (or later) it wouldn't have become so laden with trench warfare. Certainly science and technology were considered essential as regards breaking the resultant deadlock.

The Allies unsuccessful attempt at supporting Russia, via the Turkish Dardanelles Straits, floundered in early 1916. This set-back forced Britain and France to intensify their efforts on the Western Front. The ensuing Battle of the Somme, several months later, turned into *the* bloodiest battle in world history, with more than one million casualties (few strategic gains were made). Such colossal gestures contrasting sharply with the actions of just one enigmatic British soldier, namely T.E. Lawrence (better known as Lawrence of Arabia), who helped to initiate a successful Arab Revolt against the Turks (1917-18). Lawrence's contribution subsequently provided for both Britain and France expanding their empires in the wake of hostilities.

In 1917, as unsupported Russia collapsed into terminal revolution, America came to the assistance of both Britain and France. The combined weight of the English-speaking world and their French allies – plus the more effective use of aircraft, tanks and artillery – led to Germany's eventual capitulation. Germany unconditionally surrendering to the Allies on the 11th day of November 1918 (i.e. Armistice Day). The subsequent Treaty of Versailles then provided for the future containment of Germany (or at least that's what was anticipated). Of its many articles, it was article 231 which expressly blamed Germany for the war, and which imposed severe reparation payments – payments which led to hardship and resentment amongst many Germans.

In the First World War's immediate aftermath it appeared as though the purportedly peace-keeping League of Nations would serve science remarkably well, providing a much better environment in which to crack nature's deepest codes and realize its most significant powers. It was within this new climate that science once again became dominated by physicists from all across Europe. But as scientists began to seek out a new range of conceptual devices, to both illustrate and explain natural phenomena, the true nature of man began to cast a dark shadow over events. The paradox prevailing at this time was that hypothetico-deductive research and the scientific method *were* amassing insights, but not about man's true nature. That, it transpired, would take history.

- ## Science imitating nature?

In 1900 German physicist Max Planck (1858-1947) introduced the world to the elementary **quantum of action**, i.e. the notion that light is absorbed and emitted in discrete amounts (collectively called quanta). German-born physicist Albert Einstein (1879-1955) then produced a **Nobel Prize** winning paper on the **photoelectric effect** (1905), which makes explicit reference to light-quanta (what we'd now term photons). Significantly, Einstein's paper has particles interacting with other particles, specifically the photon and **electron** – which in some ways echoed the earlier marriage of electromagnetic waves and electrical currents, but in a new and exciting way. Wave-particle duality and particle physics were therefore given a much-needed boost, thanks to Einstein's landmark work.

The subsequent escalation in particle physics (starting with the photon, electron, **proton** and **neutron**) whilst fundamentally valid, nonetheless denudes the subject of some of its more profound components. Reducing the subject in this way does, however, provide for a certain amount of secrecy as regards the brain's telepathic potential. As brain waves were discovered just two years after James Clerk Maxwell's renowned treatise on electricity and magnetism, and as the resulting conceptual framework made it much easier for scientists to conceive of **telepathy** than today, remotely-induced effects may have been conceived of prior to the quantum era – and that means prior to the First World War. Between 1900 and 1954 some of the most important concepts in physics and chemistry were developed. The **atom** (a particle with the characteristics of a given element) was shown to be electrically-neutral because it contains equal numbers of protons and electrons. The positively-charged protons, it was deduced, were located within a tiny **nucleus**, usually with the same number of electrically neutral neutrons. This is when things really took-off, for it was then determined that the negatively-charged electrons arrange themselves within **shells**, containing specific **orbitals**, surrounding the nucleus. And that these orbiting electrons could jump-up into a higher orbital, when absorbing a photon; or fall back into a lower orbital, when emitting the same. With the outermost electrons, in what is termed the **valence shell**, determining the atom's **reactivity**, i.e. its ability to bond with other atoms, creating **molecules**.

Much of the 1920s and 30s were spent arguing about these fundamental realities, prior to a final model being tentatively agreed upon. Significant among all these debates was the one which took place between Austrian physicist Erwin Schrödinger (1887-1961), German physicist Werner Heisenberg (1901-76) and Danish physicist Niels Bohr (1885-1962), regarding the precise nature of matter at the sub-atomic scale. In many respects this difference of opinion has parallels with the Kuhn-Popper debate, in-so-much as once one gets beyond the superficial differences one eventually finds grounds for reconciling the two camps (or, in this case, interpretations). In this instance, Schrödinger argued the case for wave mechanics, whilst Bohr and Heisenberg presented their "**Copenhagen interpretation**" of quantum theory.

In 1926, Schrödinger had arrived at a mathematical formula which described the distribution of an electron's energy in space, in other words its wavefunction. Drawing upon the ideas of French physicist Louis de Broglie (1892-1987), who suggested that matter may possess wave-like properties, Schrödinger produced his famous wave equation. Some would say that Schrödinger was trying far too hard to represent the complex realities of nature, at the expense of arriving at a practical working model of the atom, sufficient to satisfy both the student and the chemist. Ultimately, the real winners were Bohr and Heisenberg, whose Copenhagen interpretation provides for the whole of modern **particle physics**. The Bohr-Heisenberg model says that particles *do* actually exist (something Schrödinger had questioned), but that experimentation can only furnish us with information about their position in space or their momentum, but not both (known today as Heisenberg's uncertainty principle).[4] The long-debated particle, fought over ever since the time of Newton and Huygens, had been saved from destruction. Today, the well-established Copenhagen interpretation allows for the kind of physics and chemistry we safely encounter in textbooks the world over. This is health and safety at its most intangible, for these contemporary models prevent the misuse of chemistry. A student of telepathy would have been much better served by Schrödinger's arguments, rather than Bohr's and Heisenberg's. As for the world, it was much better served by the latter.

Ch2: Core sciences

• Cathode rays and sub-atomic structure

Back in 1808, the English scientist John Dalton (1766-1844) published a book entitled A New System of Chemical Philosophy, in which he stated that all things are made from atoms. Dalton's atoms appear in his notes as solid-looking circles, with various patterns and inscriptions on their surfaces, sufficient to differentiate between the known elements. Using the symbolism of Dalton, H_2O would appear as a single circle (oxygen), with two other circles, containing dots (hydrogen), stuck to its surface. Although lacking in detail, this is still the model we use today. Dalton didn't realize that water consists of hydrogen and oxygen in the proportions two to one, but implicit in his model is the notion that **compounds** are formed when atoms of various elements combine in specific proportions.

In fact, it would take many years before the atomic weights (or, more accurately, the **relative atomic mass**) of all the various elements were accurately determined; and well over a century before chemical bonding was satisfactorily explained. Even so, Russian chemist Dmitri Mendeleev (1834-1907) had noted that the chemical elements, when arranged according to their atomic weights, exhibit an apparent periodicity. And so, in 1869, with the known elements carefully arranged into rows, called **periods**, and columns, termed **groups**, Mendeleev successfully devised the world's first **periodic table**. This table had the remarkable property of predicting the existence of unknown elements, and even anticipating their chemical characteristics.

With the study of the elements very much advanced the emphasis then shifted towards the late-Victorian obsession with electromagnetism and the early 20th century pursuit of conceptual models of sub-atomic structure and chemical bonding. One of the most important clues as to sub-atomic structure came in the form of cathode rays (first observed in 1869). British physicist John Joseph Thompson (1856-1940) discovered, in 1897, that these rays comprised so many negatively-charged particles, i.e. electrons. It therefore transpired that electrical currents were able to bridge empty space. Lightning being an example of electrons making a similar journey, travelling as they do from a region possessing a superabundance of negative charge to one of significant positive **charge**.

14

These cathode rays would literally stream along the length of a glass tube containing a vacuum, provided there was a negatively-charged terminal, known as a cathode, at one end, and a positively-charged terminal, known as an anode, at the other. The electrons emitted by the negatively-charged cathode were being drawn inexorably towards the positively-charged terminal at the other end in a continuous flow (as opposite charges attract). Crucially, Thompson realized that the paths of these negatively-charged particles could be manipulated using applied electric and magnetic fields – a discovery which shed much light on the dynamics of electron behaviour.

As these negatively-charged electrons are found within electrically neutral atoms, Thompson concluded that the atom must have bound within it an equal amount of positive charge. So, in 1907, Ernest Rutherford (1871-1937) began a whole series of experiments aimed at deducing the location of this positive charge. Hans Geiger (1882-1945) and Ernest Marsden (1889-1970), working under Rutherford, fired positively-charged **alpha particles** at thin sheets of gold leaf, and watched as approximately 0.01% bounced back (the others simply streaming through to the other side). Like Rømer, 230 years earlier, Rutherford applied Ockham's razor and concluded that a minute fraction of these alpha particles were being deflected back by points of like-charge.

As for the orbiting electrons, each of these was deemed to carry a single negative charge; and it seemed probable that the nucleus contained an equal number of positive charges, sufficient to make the atom electrically neutral. Rutherford, who was working on this problem in 1919, whilst the ink was still drying on the Versailles Treaty, named these positively-charged particles protons. Clearly, if an atom gains or loses electrons an imbalance will arise, with the atom becoming an **ion** in the process. The addition of electrons produces a negatively-charged ion (called an anion) and the loss of electrons produces a positively-charged ion (termed a cation). The central nervous system is heavily reliant on ions; the most important of which are sodium (NA^+), potassium (K^+), calcium (Ca^{++}), and chloride (Cl^-).

It didn't escape people's attention that if the nuclei of atoms were composed entirely of protons, sufficient to account for all the measured mass, the nuclei would contain too many positive charges. This enabled British Physicist James Chadwick (1891-1974) to successfully prove the existence of neutrons (particles with no charge). It's actually the presence of both protons and neutrons in the nucleus of an atom which gives an element its specific weight, and hence its position in the periodic table. The atom's mass is measured in **atomic mass units** (amu), but because atoms of the same element may contain differing numbers of neutrons, the amu is usually expressed as a weighted average of all these naturally-occurring **isotopes**.

Niels Bohr, who helped to develop the Copenhagen interpretation (which provided for an ever expanding list of particles), submitted a paper in 1913 entitled On the Constitution of Atoms and Molecules, in which he argued that electrons were confined to stable orbits around the nucleus of an atom; and that as long as they remained undisturbed they wouldn't radiate any energy.[5] This implied that if the electrons were disturbed, either through the absorption or emission of photons, their orbits would be affected (see Fig. 1). Einstein had already revolutionized physics by marrying together the photon and electron in his paper on the photoelectric effect. Consequently, Bohr was able to argue that an electron either absorbs or emits a photon, and that when it does its orbit is affected.

Figure 1: The general structure of an atom

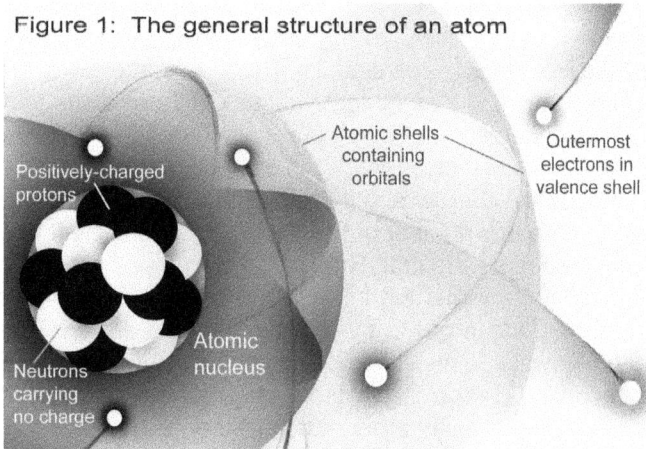

Positively-charged protons

Atomic shells containing orbitals

Outermost electrons in valence shell

Atomic nucleus

Neutrons carrying no charge

All electrons, whether streaming through a vacuum, flowing along a wire, or orbiting the nucleus of an atom, constitute moving electric charges, sufficient to indicate magnetic effects – effects which are magnified many times in certain metals (like iron), and when a current-carrying wire is coiled. Additionally, in an aerial or antennae the movement of electrons produces oscillating electric and magnetic fields which radiate out at the speed of light *and* at a given frequency. If the movement of electrons is around an electrical circuit, one might impede their flow with a bulb, producing visible light through electrical **resistance**. This then, was the perfect marriage of 19th century applied physics and the emergent theoretical physics of the early 20th century – the very moment of conception, as regards the modern world.

Much work still needed to be done in order to understand the true nature of these orbitals, and how the electrons actually arrange themselves at the sub-atomic scale. Undoubtedly, it was hoped that the distribution of these electrons would shed light on the processes driving chemical bonding, and the complex range of molecules that invariably gives rise to. But, in spite of the uncertainties, Bohr had established the first principles; i.e. that electrons jump-up and down between orbitals, depending on whether energy is being absorbed or emitted, and that how much energy an electron possesses corresponds closely to its distance from the nucleus. Modern science was starting to take shape.

• **Quantized energy and chemical reactions**

By the 1920s and 30s the search was on for a comprehensive model of the atom, one which would usefully explain the whole range of chemical reactions found in nature. Almost unbelievably, that task was more or less complete within a decade, thanks to some of the biggest names in the history of science. First came the collaboration of Austrian physicist Wolfgang Pauli (1900-58) and Danish physicist Niels Bohr on the **Aufbau principle** (1920-24). Then, Pauli swiftly added to this his very own **Pauli exclusion principle** (1924). And, finally, came **Hund's rule** (1927), named after the German physicist Friedrich Hermann Hund (1896-1997); thereby completing this triptych of closely-related models. No sooner had the ink dried on these particular papers than the world economy shook due to the Wall Street Crash of 1929 (an event which frustrated German scientific funding and initiative through its effects on the global economy).

These three essential edicts – the Aufbau principle, the Pauli exclusion principle and Hund's rule – underpin the whole of modern chemistry, and add fundamentally to our understanding of quantum physics. By the time Adolf Hitler (1889-1945) became chancellor of Germany, in 1933, following the economic disarray of the preceding few years, German, Danish and Austrian scientists were all fine-tuning the paradigm that is modern chemistry. But, whilst many scientists distanced themselves from Hitler's Third Reich, that wasn't the case with Werner Heisenberg, who became the director of the Kaiser Wilhelm Institute in Berlin, between 1941 and

1945. The youthful Heisenberg had, by that time, met and argued with some of the biggest names in science. Conceivably, those names had helped him to appropriate the paradigm we know today. The current paradigm, which is heavily-rooted in the physics and chemistry of the early-20th century, is of immense practical importance, but nonetheless devoid of certain key insights. By understanding the current orthodoxy, and then contrasting it with much deeper realities, the reader begins to comprehend the sheer magnitude of the stakes fought over so tenaciously by the Allies in those years. With this in mind, let's return to our deconstruction of the atom, and science's attempts at making sense of that cloud of orbiting electrons, so critical in driving the modern world.

Whenever one tries to imagine an atom containing many electrons – for example, oxygen, which has eight electrons – one is confronted with the question of how those particles actually arrange themselves in space. This is no easy task, as one must derive from such a model how this distribution of 'quantized energy' provides for every conceivable chemical reaction. In the case of oxygen, the oxygen atom bonds readily to two hydrogen atoms to form water. But, why? What state exists inside the oxygen atom that makes this not only possible, but universally true for all atoms of oxygen across the entire universe? Ultimately, it's about energy; the amount, its distribution, and whether it can be shared.

The Aufbau principle, or *Aufbauprinzip* (which translates from the original German as 'building-up principle') simply states that as one works through the entire periodic table of elements the electrons fill the atomic orbitals sequentially, starting with the lowest energy orbitals first. The lowest energy orbitals are adjacent to the nucleus, and the highest energy orbitals are farthest away. Some of these orbitals are closer together than others, giving rise to naturally-occurring shells and subshells. So, for example, the orbital in shell 1 (which is closest to the nucleus) has to be filled prior to the higher energy orbitals in shell 2, and so on.

Not all shells have the same number of orbitals (see Fig. 2). Shell 1, for example, has just one 's' orbital. Whereas, shell 2 has a total of four orbitals, comprising one 's' orbital and three 'p' orbitals (divided between two subshells). Because each orbital can only accommodate two electrons (due to the Pauli exclusion principle), shell 2 can only accommodate eight electrons in total. Additionally, Hund's rule states that where there are several orbitals (in a given subshell), each one must be filled with one electron, prior to the addition of a second. The reason for Hund's rule is that electrons are like-charges, and are inclined to repel one another.[6] So the lowest energy state is achieved through partial filling first. Figure 2 illustrates the distribution of electrons in an atom of oxygen, using these three basic rules, i.e. its **electronic configuration**.

Figure 2: The electronic configuration of an oxygen atom

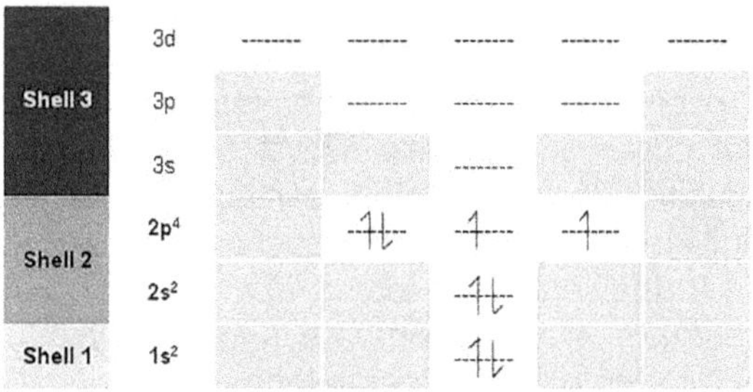

The electronic configuration of oxygen is $1s^2$, $2s^2$, $2p^4$ (as per the above table). Oxygen is highly reactive due to the number and distribution of its outermost electrons. Note, electrons are denoted by the up and down arrows.

The Pauli exclusion principle is of profound importance in quantum physics. Pauli assigned to the electron four **quantum numbers**, three of which pertain to its position in space – the fourth of which relates to the electron's so-called **spin** (of which there are two kinds). Pauli postulated that no electron with the same four quantum numbers could occupy the same orbital. So, although electrons could overlap spatially, only two could ever occupy the same orbital (and only then because they had opposite spin). Whether these quantum numbers fairly reflect the most elementary laws of the cosmos, or are simply a mathematical convenience, is an argument best left for physicists.

Of far greater practical importance were the ideas of the American chemists Gilbert Lewis (1875-1946) and Linus C. Pauling (1901-94), and the contributions made by German physicists Walter Heitler (1904-81) and Fritz London (1900-54); who, together, did much to revolutionize our understanding of chemistry through their work on valence bonding. The outermost valence shell is the one containing the electrons which participate in chemical bonding. The relative number of these **valence electrons** heavily influences the atom's overall **electronegativity** value, i.e. its tendency to attract electrons, and therefore to bond with other atoms. The magnitude of the difference between electronegativity values determines the type of bond which occurs – whether it be **covalent bonding** or **ionic bonding**. In all cases, however, the principal aim of such

19

bonding is the generation of full valence shells – which may involve surrendering electrons or acquiring them.

By the outbreak of World War 2 the whole foundations of modern chemistry had been firmly established. The contribution made by German scientists simply fueled the conceit of German nationalists, and further added to the rhetoric of National Socialists within the Nazi Party. German belligerence and imperial ambition coincided with a growing awareness of the brain's telepathic potential amongst its chief opponents. Germany's adversaries would continue to realize that potential in secret, whilst containing their enemy's imperial ambition through military conflict. The English-speaking world wasn't about to divest itself of such secrets until the extremes of **Left** and **Right** were spent forces. In the century which followed the true nature of man was laid bare – not by psychoanalysis, but by history, with the political **Centre** thoroughly educated as to the hazards posed by the same.

· Victory, in spite of all terror

The interwar years (1918-39) had witnessed the rise of Adolf Hitler's Nazi Party, whose fascist agenda and empire-building led inexorably to World War 2. When one thinks of Nazi Germany, one thinks of militarism, unbalanced budgets, anti-democratic government and political oppression. Conversely, in Great Britain and America the reverse was happening, the burden of the First World War leading, if anything, to a scaling down of the military, a return to more balanced budgeting, and the extending of the franchise, or vote, to women of full-age. It was in these years that the British Empire reached its greatest extent, due to the acquisition of German and Turkish-held territories, before fading gradually into an association of equal partners, termed the Commonwealth. This slow decline being driven, in part, by cut-backs in military spending and the return to more balanced budgets.

It was therefore with great reluctance that Great Britain declared war on Germany, in September 1939, following the latter's invasion of Poland (Austria and Czechoslovakia having, by then, already fallen to the Nazis). Then, in 1940, Denmark, Norway, Belgium and the Netherlands all succumbed to German occupation. This German blitzkrieg, or lightning war, trapped the British Expeditionary Force, along with the French Army, on the north coast of France, at Dunkirk, where they were evacuated to the south coast of England by a flotilla of naval and civilian vessels. Successful though this evacuation was it nonetheless signalled the collapse of France, whereupon Great Britain became the only declared enemy of Germany. Hitler then attempted to soften-up English defences prior to a full-scale invasion of the British Isles. The resulting air-war was named the Battle of Britain, and it began in the summer of 1940. Under the leadership of its Prime Minister, Sir Winston Churchill (1874-1965), Britain rallied behind the proverbial few in the Royal Air Force, charged with the nation's defence. In spite of the loss of many aircraft and numerous bombing raids on central London (otherwise known as the Blitz), the Nazi invasion plan nonetheless collapsed. Thus, the Battle of Britain entered into English folklore as a truly historic achievement by Britain's armed forces. However, those forces couldn't hope to single-handedly defeat the entire Third Reich and its Axis allies. That, it transpired, would take several factors, some of them related to the psychology of Hitler himself.

Hitler's confused imperial ambitions had caused him to make war on Great Britain, a country he claimed to admire, having forged a non-aggression pact with those he ideologically despised, namely, the Russians. The Russians, or Soviets, having become communists, following the country's terminal collapse and subsequent revolution in 1917. Thus, imperial Russia had morphed into the **Union of Soviet Socialist Republics**, or USSR. Hitler, who was on the extreme Right of politics, had come to power with a mandate which included countering the soviet threat to the east (the Russians being on the extreme Left of politics). Contrary to his earlier agreement, and with his invasion plan of Britain in tatters, Hitler ordered his troops to invade the Soviet Union (June 1941). Thus, Russia became Britain's first major military ally of the war.

Together, Britain and Russia would gradually grind-down the German war machine, helped and assisted by Commonwealth troops, and later by the USA. Those fighting on Britain's side being termed the Allies, or Allied Forces. Facing them were the Axis Powers, who comprised Germany, Italy and Japan; all of whom were actively engaged in empire-building at this time. The Axis Powers thought that such an alliance would deter the USA from entering the war (an American attack on one would make enemies of them all). In the event, the Japanese felt compelled to attack Pearl Harbour, a US Naval Base in the state of Hawaii, forcing American involvement (December, 1941). Britain now had two major military allies – both Russia and the USA.

By the end of 1942, within 12 months of the USA entering the war, the war was beginning to turn in the Allies favour. American successes in the pacific, against Japan, being matched by British successes in North Africa against German and Italian forces. Shortly after this the rag-tag remnants of Hitler's invading army would begin slowly retreating out of Russia. As the direction of the war began to turn in the Allies favour, the Nazi's wholesale extermination of Jews – the **Holocaust** – reached its height. By the end of the war Hitler's "final solution" had led to the murder of 6 millions Jews. With evidence of such barbarity simply fuelling the argument that the war should be prosecuted, whatever the cost.

The Allies eventually invaded Italy in September 1943, and after prolonged and exhaustive fighting they succeeded in conquering it. Italy being the first of the Axis Powers to succumb to Allied occupation. The subsequent D-Day landings in Normandy (June, 1944) heralded the start of Operation Overlord, the Allied operation which liberated both France and Belgium. By December 1944, Germany was the sole remaining Axis power in Europe. Russian troops entered the German capital, Berlin, in April 1945, and discovered that Hitler had committed suicide. The 8th day of May 1945 being proclaimed as VE Day (Victory in Europe day). With Germany and Italy occupied, only Japan continued to fight on, relying increasingly on somewhat desperate kamikaze attacks against Allied warships.

Unbeknownst to the Japanese, however, scientists in America – not all of them native-born Americans – had committed to manufacturing the world's very first atomic bomb. But to achieve that goal western science needed to add to its broader understanding of the atom, and it would need to do so in an environment increasingly governed by political and military considerations. What the scientific method fails to stress is the importance of applied physics and chemistry, as regards the identification of problems. As opposed to observation and research based solely on the prevailing paradigm. In that respect, America's open-minded pragmatism was beginning to furbish it with an unprecedented destructive capability.

•   Mass-energy

Free protons are simply the nuclei of hydrogen atoms – in other words, single protons. Alpha particles are helium nuclei – comprising two protons and two neutrons. Whilst negatively charged **beta particles** are high energy electrons, i.e. cathode rays. Everything which is made of atoms can be broken-down into these constituent parts – and both hydrogen and helium atoms could easily arise from these simple ingredients. But to produce heavier elements takes something more dramatic. Hydrogen atoms tend to pair-up, achieving full valence shells in the process – the resulting hydrogen molecules being the simplest of all molecules (typically written as H_2). If all that existed *were* hydrogen molecules and helium atoms things would be chemically rather inert. And that, as they say, would be that; but for one vital ingredient – gravity.

Gravity is the force which produces the materials the **big bang** forgot to produce, and which are so vital to complex life.[7] So, gravity provides for many things, but could it bring together free protons, alpha particles and beta particles sufficient to create a black body? Something which is a perfect absorber and emitter in all wavelengths. In theory, we'd require gravity to produce every conceivable chemical element, sufficient to leave no dark bands or lines in our visible spectrum – and, indeed, no absent wavelengths of any kind beyond the visible. For gravity to create such an object requires, at the very least, **fusion** within stars – but, additionally, the supernovae which gravitational collapse occasionally gives rise to.

Only gravity is able to squeeze together these ingredients under conditions of immense heat and pressure, as found at the sun's core, sufficient to produce, for example, a carbon nucleus – any surplus mass radiating out as pure energy (experienced on the earth as warmth and sunshine). That nucleus of carbon may one day attract to itself six electrons from out of the surrounding plasma, producing an electrically neutral carbon atom in the process. If so, those six electrons will obligingly fill the available shells and orbitals sequentially, with **electron jumps** providing for the absorption and emission of photons in wavelengths specific to that element. The continued compression of such nuclei, into a whole series of chemical elements, adding incrementally to the overall spectroscopic complexity of the cosmos.

Gravity therefore concentrates positive charges with a periodicity bordering on the tabular – one need only consult the periodic table to see evidence of the same. This naturally gives rise to atoms possessing various chemical properties and unique spectral characteristics. These atoms are electrically neutral, and if left completely alone the energy they contain remains constant. Even if an atom gains or loses energy, due to the absorption or emission of photons, it remains electrically neutral. The ineluctable conclusion is that photon's carry no charge. But is the force

23

mediating between electrically-neutral atoms, gravity – gravity being essential to atomic interaction? Even with the benefit of gravity producing a rich variety of elements, it's still difficult to produce the idealized black body. This is where chemical bonding comes in. Covalent bonding, especially, affects the absorption and emission spectra of all the molecules it produces. These molecules further add to the list of recordable spectra (invariably in wavelengths beyond the visible). In order to continue adding to this list we need to be really quite cunning, creating large covalently bonded macro-molecules, surrounded by others of varying complexity, and all linked together via a rich variety of forces. A process which would result in some hitherto unknown spectra.

But even this wouldn't furnish us with every conceivable wavelength, and we might have to create ever heavier elements in the laboratory, which are inherently unstable, just long enough to emit some exotic frequency, or promote mobile phone technology in the vain hope that one day all the recordable spectra will be accounted for. Armed with a box of free protons, alpha particles and beta particles, one might forego this challenge of trying to create an elusive black body, and make do with creating intelligent life instead. Not only is complex neurology easier to craft, it also has bound up with it a naturally-occurring telepathic potential, that's relatively easy to exploit.

It's also easier to exploit the destructive power of uranium than split an alpha particle through **fission**. Alpha particles have an extremely high binding energy that the mutual repulsion of the two positive charges would find impossible to overcome, even with the benefit of additional neutrons. This isn't the case with much heavier, unstable elements, like uranium. Uranium will naturally transmute into a lighter element, called thorium, through alpha decay (spitting out a helium nucleus as it does so). It can also be cleaved into lighter elements, called fission products, releasing immense amounts of energy in the process. This release of energy can either be arrived at slowly under very controlled conditions, as in a nuclear power station, or it can be explosive.

This destructive potential was seized upon by those scientists who worked for America's Manhattan Project, whose stated aim was to produce the world's first atomic bomb. Some of these scientists were individuals estranged from their own countries by the rise of fascism in Europe. America's leading nuclear physicist Robert Oppenheimer (1904-67) became the scientific director of the project's laboratory at Los Alamos, New Mexico, between 1943 and 1945. This is where the 'Little Boy' and 'Fat Man' fission bombs were developed; the ones which laid waste to the Japanese cities of Hiroshima and Nagasaki, in August 1945. The Japanese, of course, capitulated unconditionally.

Assisting Oppenheimer in those years were physicists such as Richard Feynman (1918-88), Danish born Niels Bohr, German born Hans Bethe (1906-2005), and Italian born Enrico Fermi (1901-54). The Manhattan Project took the science of secrecy to a whole new level, and left in its wake a whole new generation of intrigues, custom built to obscure some of nature's most dangerous realities. As for Heisenberg, he was still struggling with our present paradigm; the orthodoxy which **mainstream** science has inherited, and which makes telepathy almost impossible to deduce.

•   Quantum subterfuge

Modern chemistry may have equipped itself with ideas relating to electronegativity, valency and the distribution of all those electrons, but in the wake of hostilities physics took it upon itself to promote a whole new science: namely, the science of **quantum electrodynamics**, or QED. QED was exclusively about the interaction of light and matter; or, if you prefer, the interaction of photons and electrons. In many ways it wasn't a new science at all, as the preamble to it included Maxwell's famous equations (1873), Hertz's radio transmissions (1887), Planck's discovery of the quantum (1900), Einstein's photoelectric effect (1905), and Bohr's paper on the constitution of atoms and molecules (1913), not to mention numerous other insights and additions. No, this wasn't a new science at all, it was simply the appropriation of that whole subject by the political and military powers.

Quantum electrodynamics was rooted in quantum physics, and if one marries quantum physics to the Copenhagen interpretation then all those quanta comprise so many particles. A principle advocate of particle physics at this time was Richard Feynman. Born in Far Rockaway, New York, USA, in 1918, Feynman gained a degree in physics from the Massachusetts Institute of Technology, prior to undertaking his PhD at Princeton. Whilst at Princeton he was headhunted for the Manhattan project. After which he taught theoretical physics at Cornell University,

New York (1945-50), alongside his friend and associate from Los Alamos, Hans Bethe.

It was during his time at Cornell that Feynman presented his ideas about QED to a number of eminent physicists, most significantly at the Shelter Island Conference (1947), the Pocono Conference (1948), and the Oldstone Conference (1949). Ostensibly, Feynman would argue that his reasoning was motivated by the need to inculcate clearly, whilst reducing the mathematical infinities. But this is to believe that the Soviets weren't interfering with surface traffic between West Germany and that part of Berlin controlled by the Western Allies – in the first of several gestures which would herald the start of the **Cold War**. The Cold War being a state of political and military tension arising between the **Western Allies** (namely, USA, Great Britain and their allies) and the **USSR**, following World War 2.

In June 1948, amid Feynman's presentations, the Western Allies announced the creation of West Germany (i.e. the German Federal Republic). The following year, in October 1949, just 5 weeks after successfully exploding their first atomic bomb, the Soviets announced the creation of East Germany (i.e. the German Democratic Republic). As for the German capital, Berlin, that became permanently divided between East and West, necessitating airlifts; all of which culminated in the building of the **Berlin Wall**, by the Russians. Thus Germany, Europe and the wider world became divided, not simply along ideological lines, but also by a full-blown nuclear arms race. And so, it was within this climate of escalating tension that Feynman gave his presentations and argued his case. And his case, it transpired, was somewhat disingenuous.

Feynman began by arguing that particles might be said to communicate all their influence (e.g. mass, charge and energy) by means of other particles, and that one might choose to ignore rather abstract fields. In the words of Piers Bizony, author of the book Atom (Icon Books, 2008) "Feynman wanted to cut through the infinities, and eliminate the complexities of talking about fields, by reducing electron behaviour to just three basic ideas. 1) An electron moves through space. 2) A photon moves through space. 3) An electron either emits or absorbs a photon."[8] In some respects Feynman was seeking to pacify those who'd noticed that not all variables were neatly quantized, and that infinities were still capable of plaguing their mathematical calculations.

Figure 3: An example of a Feynman Diagram

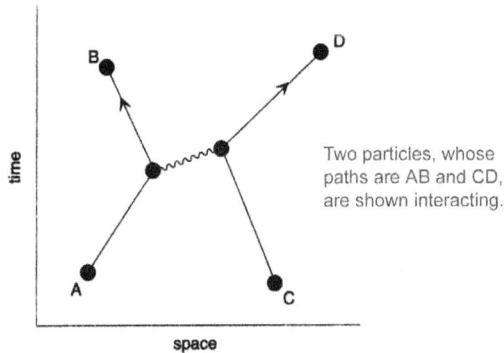

Two particles, whose paths are AB and CD, are shown interacting.

Also, chemistry at this time was progressing rapidly through its use of some rather simple graphical devices (related to the Aufbau principle, Pauli exclusion principle and Hund's rule), and perhaps Feynman was seeking to emulate those relatively undemanding models, while still providing for so much secrecy. To that end, Feynman began to rely increasingly on some rather stark diagrammatic representations, showing the interactions between various elementary particles. These so-called **Feynman diagrams** (see Fig. 3) found favour with the English-born physicist Freeman Dyson (1923-2020), who became a close friend and acquaintance of the three Los Alamos veterans, Richard Feynman, Hans Bethe and Robert Oppenheimer.[9] Dyson was convinced that Feynman diagrams were of enormous scientific value, and with heavyweight support from the veterans of the Manhattan Project, including Oppenheimer himself, they quickly became established.

Subsequently hyped as the most precise and successful theory in the whole of physics it all seemed pretty much sewn-up in Feynman's favour, with America the final arbiter of the truth as regards QED. So, America was now well-placed to promulgate the most successful theory in physics to a world deeply divided along ideological lines. But a lingering question remains, were these giants of political and military intrigue being entirely honest at such a decisive moment in modern history, with the dust scarcely settled on the Russian's first atomic bomb test? The truth is that the deeper realities behind QED were to be veiled in the utmost secrecy, and subsequently deployed in the wider strategic interests of the Western Allies.

America had certainly furbished the world with a compelling conceptual device, but many of those missing details were actually far more important than was professed to at the time – some of which relate to the influence of electric and magnetic fields. Fields which were known to affect electron behaviour, and in all probability the absorption and emission of photons. By downplaying the significance of such fields a faction within American science had made it far more difficult to deduce the existence of the human brain's naturally extant telepathic potential. But to comprehend why that potential exists at all, one must first understand why **destructive interference** makes it necessary to possess that most latent of faculties.

•   Destructive Interference

Where the crest of one wave and the trough of another meet, both having the same amplitude, frequency and wavelength, they destructively interfere with one another. When this happens the waves are said to be exactly out-of-phase.[10] If an absence of such interference is biologically essential, for whatever reason, its presence could have pathological implications. Expressed more simply, if human neurology evolved to avoid this type of interference then the imposition of such frequencies could have profound implications for the person. Earlier in this chapter, in the section entitled mass-energy, it was argued that it isn't possible to create an idealized black body, and spoke of the impossibility of filling-in all the available gaps in the electromagnetic spectrum. It's as though complex neurology has evolved to fill-in those spaces left blank by an absence of radiation.

The human brain is able to avoid destructive interference by capitalizing on those gaps in the electromagnetic spectrum (see Fig. 4). But in order to do this the physical structure of the brain must differ between persons; sufficient for person-specific electric and magnetic fields to arise. Under the influence of those fields the electronic configuration becomes unique to the individual – perhaps not in terms of the shells, subshells and orbitals themselves, so much as slight modifications to the same. The manner in which this happens naturally gives rise to person-specific spectra. Those person-specific spectra then make remote manipulation, mental telepathy and the superimposition of persons possible, because nature never anticipated those wavelengths being actioned by intelligent life.

Figure 4: Electromagnetic spectrum

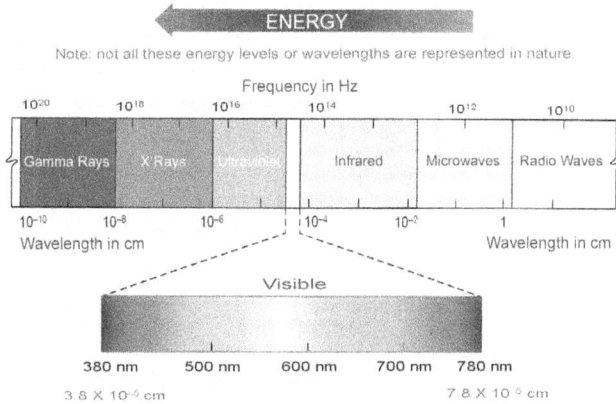

Since the 19th century people have realized that there's no intrinsic reason why electromagnetic waves couldn't be many times longer or shorter than the known spectrum. Some sources speak of a near infinite range of wavelengths – crucial, if one is to argue that people possess a unique **absorption profile**. The person's absorption profile might be expressed as a whole series of wavenumbers (a **wavenumber** being the **reciprocal** of a particular wavelength). These wavenumbers would correspond with a variety of physical, physiological and cognitive functions, pertaining to a particular person. The electronic configuration, like the human frame, can be expressed in an idealized form – but the reality is endless variation, resulting in an absorption profile unique to the individual.

Two identical waves neutralizing one another at a point, both of which are photons, is reminiscent of a photon and an anti-photon annihilating – proof that **antiparticles** really do exist! Thus, the presence of identical waves may take on the connotations of a barrage of anti-particles. However, these aren't *actual* antiparticles and the results are likely to be more mixed. Indeed, waves which are perfectly in-phase are magnified, so we mustn't make too many assumptions. What we can say is that cognition is seriously threatened if it's collapsed by background radiation or supplemented in some *ad hoc* fashion. And this is why consciousness has arisen within a radiological niche.

Radiation quite-literally radiates, hence a given light source grows dimmer as one recedes from it, as per the **inverse square**

29

law. Gravity also conforms to the inverse square law, and so one can't easily separate the two phenomena. As light is emitted as electric and magnetic waves possessing precisely the same amplitude, frequency and wavelength, the light arising from a single point doesn't normally interfere with itself; unless, that is, the light is first divided and the two separate paths are then redirected back towards one another. And this appears to be the key to understanding destructive interference under controlled conditions.

To observe destructive interference one must use light of a single colour from a single source, with the beam divided into two parts, which are then brought back together at a screen.[11] If two completely different light sources are used no stable pattern emerges. In other words, radiation arriving from two different bulbs, or brains, produces a totally incoherent result. This doesn't mean that interference isn't occurring, or couldn't occur, simply that it can't be discerned. And so it is in human neurology, no clear evidence of destructive interference presents itself, but it remains nature's hidden menace – the illusion of non-interference you might say. The most tangible proof we have, of this unseen hazard, is evolution's longstanding attempts at avoiding the same.

So, destructive interference can and does occur, even though demonstrating it can be problematical. But in order to understand precisely how nature has configured itself to avoid this surreptitious threat we need to examine structural bonding. For hydrogen and oxygen atoms to have distinct spectra, and then to bond together to create water molecules, which themselves have unique spectral properties – well, something must be happening which goes beyond mere structural concerns. The very process of bonding must be producing a compound which absorbs and emits uniquely. And so it is with the human brain.

Ch3:   **Structural bonding**

•   Bringing the pieces together

There are well over 100 billion brain cells in the average adult brain – that is to say, the total number of **neurons** and **glial cells** taken together exceeds that figure. The glial cells provide nourishment and support for the neurons, which are the primary nerve cells. Each of these brain cells comprises well-over one million molecules, with the largest organic molecules possessing several thousand million atoms – and so the total number of electrons in the average adult human brain *will* exceed 10^{18} (written as a one, followed by eighteen noughts), and in all probability the final figure is likely to be significantly higher. As all human brains are physically different the electric and magnetic fields, generated by the same, will be person-specific – due to differences in the number of electrons, their spatial distribution and the magnitude of the fields.

The **electric fields**, associated with electrons, are simply the regions of influence surrounding these negatively-charged particles, and the principal reason for Hund's rule, i.e. that electrons are like-charges and are inclined to repel one another, so that each orbital is partially filled first prior to the addition of a second electron. Far more intriguing, in terms of this debate, is the influence of all those **magnetic fields**, whose effects appear considerably more complex and profound – not least of which because together they're instrumental in the absorption of photons and indirectly associated with the movement of electrons. The orthodoxy disingenuously contending that all magnetic effects stem from moving electric charges, with an electron said to contribute a magnetic influence because none is ever stationary, whether it is confined to an atomic orbital or being shared between atoms.

These magnetic effects aren't always fully appreciated because conventional chemistry dotes on the electronic configuration of things, rather than the **magnetic configuration**. In many respects the electronic configuration reflects the distribution of the electric fields, because it denotes where the electrons are positioned, and hints at their region of influence. But that still leaves much uncertainty as to the strength and distribution of all those magnetic fields – which are inherently more complex, and inclined to amplify or weaken one another, depending on their orientation. At present, if one were to ask a student of chemistry what the electronic configuration of helium is, they'd probably answer $1s^2$ (helium having two electrons filling an 's' orbital in shell 1). Now ask them for the magnetic configuration – they'll probably be stuck for an answer.

All atoms contain equal numbers of positive and negative charge, with electrostatic forces directly responsible for holding the electrons in their respective orbits. Accordingly, one wouldn't expect the atom to attract more electrons than there are protons in the nucleus, and yet atoms like oxygen (with high electronegativity values) are hungry for electrons. The simple explanation, of course, is that the acquisition of two electrons, by an oxygen atom, results in a full valence shell. But even this fails to fully account for this deeply mysterious force of attraction. In the case of water molecules the shared electrons strongly favour the oxygen atom, resulting in a permanent separation of charge. You'd be forgiven for thinking that the valence shell fills first, producing this permanent separation of charge; but in reality the magnetic forces are filling the valence shell, via charges which have become separated.

According to this synopsis magnetism enables certain atoms to attract more electrons than they have protons; permits others to lose electrons, sufficient to produce regions of positive charge; and dictates the way in which electrons configure themselves within both atoms and molecules, sufficient to give rise to valency, chemical bonding and absorption and emission spectra. So powerful is the magnetic component, relative to the electric, that its influence can even create the illusion of shells and subshells – and, on occasions, so consolidates the electrical forces that it gives rise to noble gases and chemically-inert metals. Moreover, overlapping magnetic fields, derived from the valence shells, amplify one another – contorting the valence electrons into molecular orbits.

- **Atoms into molecules**

Just four elements make up 96.6% of the human body, by mass. These elements are oxygen (61%), carbon (23%), hydrogen (10%), and nitrogen (2.6%). Oxygen (O) is extremely chemically reactive, making it vital for **respiration** – so reactive, in fact, that metabolic reactions involving oxygen frequently produce **free radicals** which damage tissue, accelerate aging and contribute to genetic mutation. Carbon (C), which is the basis of all **organic chemistry**, provides for very large molecules, from proteins to lipids, and from carbohydrates to nucleic acids (all of which rely on carbon chains, branched carbon chains and carbon rings for their overall structure). As for hydrogen (H), that enables a very important type of chemical bond to occur, called a hydrogen bond. And finally, we have nitrogen (N), which is second only to oxygen as regards reactivity, and is found throughout the body in proteins, hormones and DNA.

A word of caution, prior to continuing, the term carbohydrate (literally meaning: *carbon-water*) implies carbon (C) and water (H_2O). Whilst plants do produce carbohydrates from carbon dioxide and water, via photosynthesis, the carbohydrates we ingest actually contain C, H and OH (hydroxyl groups). This is because water molecules can **dissociate** into H and OH, with these two bits helping to form larger structures. Therefore, *actual* water within the human body is confined to **aqueous solutions**, such as the solvent found inside cells.[12] Nevertheless, I'll persist in using H_2O as an example of structural bonding, as it's by far the easiest way to convey many of the key concepts. Incidentally, highly-reactive hydroxyl free radicals (OH), produced through dissociation, can harm tissue, because this constitutes valency at its most aggressive, i.e. the single unpaired electron in a valence shell.

The elements carbon and hydrogen are essential to the formation of macro-molecules and larger bodily structures; whilst oxygen and nitrogen assist in producing the chemical reactions needed to sustain life. The remaining 3.4% of the human body (by mass) includes the kind of elements that you might anticipate finding, e.g. calcium, iron and sodium, together with a number of trace elements, e.g. selenium, molybdenum and vanadium. These trace elements are toxic in excess, but their absence can severely frustrate life: for example, selenium, which is light-sensitive, is important in vision; while a lack of freely-available molybdenum is thought to have heavily constrained the evolution of early complex life. As for vanadium, its role is still hotly debated. But one thing is for certain, without these trace elements all that exists would exist very simply.

The electronic configurations of the four principal elements found in the human body (namely, those of oxygen, carbon, hydrogen and nitrogen) show how many electrons are present in each atom – simply count-up the superscripted numbers in the following: O ($1s^2 2s^2 2p^4$), C ($1s^2 2s^2 2p^2$), H ($1s^1$), and N ($1s^2 2s^2 2p^3$). So, oxygen has eight electrons, carbon has six electrons, hydrogen just one, and nitrogen seven. The large numbers denote shells; so, with the exception of hydrogen, the other three elements have sufficient electrons to partially fill shell 2. This leaves only the lower-case letters, which are the orbitals themselves (shell 1 has one 's' orbital; while shell 2 has one 's' orbital and three 'p' orbitals). These configurations support or give rise to electronegativity values of 3.5 (oxygen), 2.5 (carbon), 2.1 (hydrogen), and 3.0 (nitrogen). What is immediately apparent is how much more reactive nitrogen and oxygen appear, compared to carbon and hydrogen. In all cases, partially-filled valence shells drive chemical reactions and provide for atoms bonding into molecules; it's just that some elements are more reactive than others. If we illustrate the electronic configuration of hydrogen and oxygen visually (see Fig. 5), we can see that the principal aim of chemical bond formation is to generate full valence shells – and that by achieving full valence shells, order (as they say) is restored.[13]

Figure 5: H_2O – the generation of full valence shells

A highly-reactive oxygen atom may covalently bond with two atoms of hydrogen; with all three then achieving full valence shells.

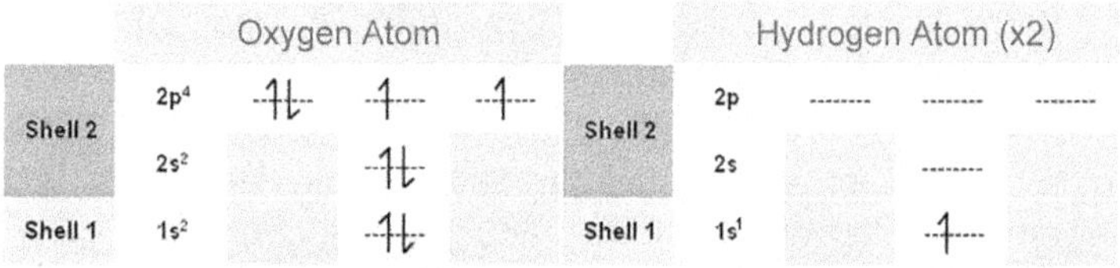

Now we can say with certainty what state exists inside atoms of oxygen sufficient to make the formation of water molecules a cosmic certainty, wherever and whenever the conditions and ingredients exist. This type of bonding is called covalent bonding, which is the principal type of bonding found in the human body. Covalent bonding frequently gives rise to **dipolar molecules**, or permanent dipoles, because the shared electrons often favour one of the bonded atoms. These dipolar molecules then have regions carrying a partial positive charge and regions carrying a partial negative charge, because the shared electrons are drawn to the atom with the highest electronegativity value.

In the case of H_2O, which is occasionally written H-O-H, the shared electrons are drawn towards the oxygen atom, and away from the two hydrogen atoms. The difference in electronegativity between the hydrogen and oxygen atoms is 1.4, so a covalent bond forms – had the difference been greater than 1.7 a different type of bond would have arisen, called an ionic bond. Ionic bonding takes place between sodium and chlorine atoms in sodium chloride (better known as salt), with the human body's only known ionic forces being called salt bridges.

Polar compounds, like water, are known to have higher boiling and melting points than non-polar compounds (a reflection of how dipoles behave when stuck together *en-masse* by intermolecular forces). This permanent separation of charge produces what is known technically as a **dipole moment** – dipole being the separation of charge and moment referring to the turning effects of a force (torque, in other words). That torque is a reflection of the **magnetic moment** and the forces exerted indirectly on the electrons by the surrounding magnetic fields. Together, these moments, which derive from the electric and magnetic fields, exert a powerful influence over the electrons, sufficient to make all those electronic configurations seem overly generic – certainly at the molecular level of organization and higher.

- **More complex attachments**

Carbon makes complex life possible by providing for a wealth of biological molecules. This chemically unique element can form into chains, rings and branches; structures which then support a variety of other elements, often in what are known as functional groups. The resulting macro-molecules include proteins (which incorporate amino acids); lipids (which are fat-like hydrocarbons); carbohydrates (such as glucose); and the all-important nucleic acids (most notably deoxyribonucleic acid, or DNA). Unparalleled molecular diversity, married to some rather ubiquitous chemistry, enables carbon-based organic molecules to generate an incredibly rich array of flora and fauna.

Consequently we find that molecules, like the elements themselves, exhibit a certain periodicity, or group in certain ways. Thus, a whole range of molecules share similar characteristics – making them, as in the above example, carbohydrates, lipids, proteins or nucleic acids. How these various types of molecule then interrelate, to form cells, organs, connective tissue and people, is down to a variety of special forces. These forces are able to assemble these molecules in a multitude of ways – ways which are persistently dynamic (i.e. living). Life therefore comprises so many semi-permanent carbon-based organisms which interact with both themselves and their environment, in an imperfectly self-replicating manner.

That process of assembly, interaction and renewal is mediated by the strength and permanency of all the various forces holding the person together. Some of these forces are weak and transient, others strong and semi-permanent; some water-loving, others water-hating; in an arrangement which plays on all those differences. Take, for example, **dispersion forces**; which are generally short-lived, very weak, and operate over very short distances. Dispersion forces arise when two molecules are in close proximity, and the electrons in one molecule repel the electrons in the other – as like charges are said to repel – sufficient to create, in the other, an **induced dipole**. The cumulative effect of dispersion forces is what makes them significant throughout the human body – for example, where cells meet (or the structures they contain).

Far more significant in terms of strength and permanency is hydrogen bonding (more commonly called **H-bonding**), which exploits the electrostatic attraction which exists between a hydrogen atom on one dipolar molecule and a particularly electronegative atom on another dipolar molecule. The typical energy associated with H-bonding is 20 Kj mol^{-1} (compared with 150-1000 Kj mol^{-1} in the case of covalent bonding). Even the strongest intermolecular forces cannot compete, therefore, with the sheer magnitude of intramolecular forces (which are generally more than ten times greater), reflecting the structural integrity of the molecule, so vital to our story. Although much weaker, the forces binding molecules will nonetheless subtly affect the magnetic forces acting on individual electrons – providing for person-specific spectra in the process.

Figure 6: Intermolecular forces

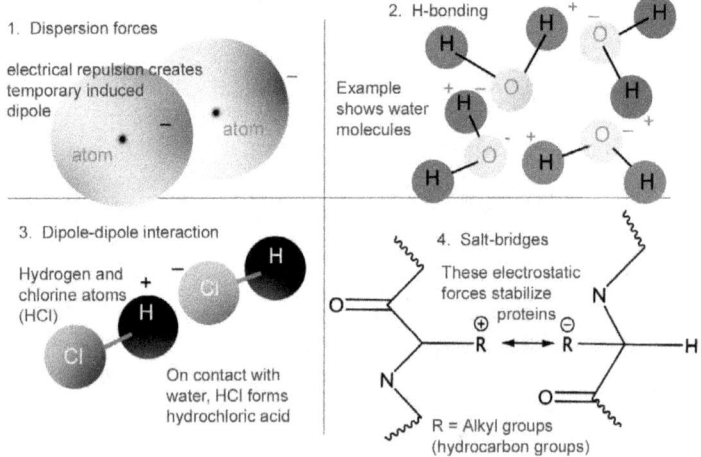

1. Dispersion forces

electrical repulsion creates temporary induced dipole

atom

2. H-bonding

Example shows water molecules

3. Dipole-dipole interaction

Hydrogen and chlorine atoms (HCl)

On contact with water, HCl forms hydrochloric acid

4. Salt-bridges

These electrostatic forces stabilize proteins

R = Alkyl groups (hydrocarbon groups)

H-bonding plays a significant role as regards the structure of proteins, and hence adds to the body's anatomical strength and three-dimensional form. Also, hydrogen bonding plays an important role as regards nucleic acids, helping DNA to adopt and retain its double-helix shape, whilst providing for its replication by not being inordinately strong (this characteristic makes H-bonding crucially important in all biological systems). Additionally, for a substance to be water soluble, for example within the aqueous environment of a cell, it has to be able to form H-bonds with the water molecules. If it can, it's **hydrophilic**, but if it can't, it's **hydrophobic**. DNA has, for example, both hydrophilic and hydrophobic components. In fact, DNA's classic double-helix shape serves to expose its water-loving parts to the aqueous environment of the cell, whilst actively shielding its water-hating components.

A similar effect can be seen in the membranes that form the walls of cells, and which consist of **lipid bilayers**. Lipid bilayers comprise a water-hating hydrophobic layer sandwiched between two water-loving hydrophilic layers. So, the inside of the membrane loves the aqueous environment of the cell, and the outside of the membrane loves the tissue fluid surrounding the cell, but sandwiched in-between is a water-hating layer. Therefore, being hydrophilic or hydrophobic adds an additional twist to intermolecular interactions. Being hydrophobic means not being able to form H-bonds with water, implying that weaker **dipole-dipole interactions** may still be possible (see Fig 6).

Finally, one must mention ionic forces, which are a common feature of some covalently-bonded molecules possessing regions carrying a full positive charge or full negative charge; a situation most often found in proteins. The ionic forces which operate between two oppositely-charged amino acid side chains, in a protein molecule, are called **salt bridges**. These ionic forces perform a vital function in stabilizing the 3-dimensional structure of proteins. Salt bridges also arise in the protein molecules making-up the sodium ion channel openings found in the membranes of neurons – and therefore help to provide for the transmission of nerve impulses, by supporting its chief apparatus.

If we examine the internal structure of a nerve cell (see Fig. 7) we see that in keeping with cells generally the neuron contains a central **nucleus**, surrounded by a cell body (which is enclosed within an outer membrane). Unlike other cells, however, the neuron has evolved to convey nerve impulses across the nervous system – both the central nervous system (comprising the brain and spinal cord) and the peripheral nervous system (comprising all those parts extending beyond the central nervous system to the extremities and sensory organs). The highly specialized semi-permeable **plasma membrane**, enclosing the nerve cell, contains **sodium ion channels**, which provide for the free-movement of positively-charged sodium ions (Na^+) into the neuron.

Figure 7: Section through a neuron

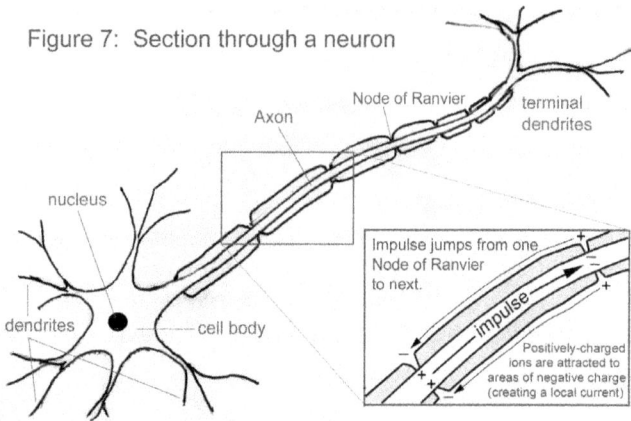

When these sodium ion channels are activated the inside of the membrane becomes flooded with positive charge, and this positive charge is able to activate adjacent ion channels. In this way, a nerve impulses is conveyed along the entire length of the neuron. Because each neuron is a discrete unit, separated by a small gap (called a synapse), the effect is one of contiguity, rather than continuity. Contiguity means adjacent to, but separate from. The chemicals which bridge the intervening synaptic cleft provide flexibility, enabling impulses to be propagated at immense speed or suppressed very rapidly. But how is this system, based on contiguous parts, remotely accessed? That's a question we now examine in a little more detail.

• **Looking beyond the electronic**

If we were to irradiate a molecule in a neuron, the whole neuron, or indeed, the entire brain itself, with many different wavelengths, some of those wavelengths would be absorbed – the resulting absorption spectrum being person-specific; most notably in the longer wavelengths, beyond the visible, due to the influence of personal biomagnetism. Even if only a few molecules in a given neuron were particularly suited to absorbing energy in a given wavenumber, it would still be possible to get energy into the said neuron, and the effects wouldn't always be merely thermal, i.e. a rise in temperature. Some of that absorbed energy would be

transduced in such a way that it causes sodium ion channels to open, creating nerve impulses, etc.

Such transduction commonly occurs when light strikes the back of the eye, and is converted into a nerve impulse; but, unlike the eye, the brain has evolved to avoid this happening in some *ad hoc* or accidental fashion, i.e. incident radiation doesn't give rise to cognitive activity, other than through the impact it has on one's senses. However, the brain is capable of irradiating itself, often in an impromptu manner, creating the sense that the brain is a casual observer of both its surroundings and itself (if not directly, then indirectly). By this means, signals emitted within the brain can easily develop into a life-long feedback of thought, memory, cognition and experience, through their immediate re-absorption.

The human brain is the most complex object in the known universe, and it has evolved to be the universe's most complex object in order to exploit those regions of the electromagnetic spectrum which have yet to be spoken for. This would explain why it possesses many more brain cells than it actually needs. The person-specific absorption profile (which personal biomagnetism naturally gives rise to) isolates the human mind, but not in an altogether inaccessible way. Telepathy is therefore artificially achievable, precisely because it's inconceivable in nature.

•   Magnetic configurations

Elementary physics tells us that where two or more magnetic effects overlap they strengthen or weaken one another, much depends on the direction of the fields. This characteristic is employed to dynamic effect in a **solenoid**, which is a simple device for converting electrical energy into mechanical thrust, by way of magnetic fields. In essence, the orthodoxy contends that an electrical current, running through a wire, produces magnetic fields – and that those magnetic fields are amplified by coiling the wire, thereby producing a mechanical force on a moveable iron core.[14] This handy little gadget (which is used as an operating switch or circuit-breaker) is demonstrative of the relationship which exists between electrons, electric fields and magnetic forces.

Even if the current is fixed, the magnetic field can still be varied, sufficient to affect the magnitude of the mechanical force, simply by altering the way the wire is coiled. And so it is with atoms and molecules; the electrons – so the orthodoxy goes – generate magnetic fields, which produce a torque effect, or moment, comparable to the mechanical thrust in the solenoid. And, as per

39

the right-hand rule, this mechanical thrust, and the magnetic fields producing it, are both at right-angles to the direction of the current (or, in this case, electrons). As the magnetic configuration is subject to endless variation – due to the strengthening and weakening of overlapping fields – electron levels and orbitals are rendered unique.

These cumulative additions and subtractions to personal biomagnetism are a consequence of factors as diverse as the type and distribution of all the various atoms; the chemical composition of the molecules they form into; the location of all the various points of charge; the angle of rotation of the said molecules; the range of electronegativity values involved; and the direction and magnitude of all the various lines of force not already accounted for above. With the sum total of all these factors being a person-specific absorption profile – one which can be tapped into remotely, due to the electrons configuring themselves in ways which provide for the absorption and emission of wavelengths and frequencies exclusive to the individual.

Twenty-five elements combine to produce this enormous variation, absorbing some visible wavelengths in the process, whilst reflecting others (giving you your visible appearance when bathed in sunlight). Personal biomagnetism simply adds to the plethora of absorbable and recordable spectra, but at much longer wavelengths – and in ways which are individual and unique. Life-forms assemble these fields and lines of force in ways which are specific to themselves, producing exclusive electron jumps in the process. The fact that each electron exists in 3-dimensional space, distinct from every other electron, hints at a discrete separation of all the factors holding it in that position. Conceivably, we may find that the electric and magnetic are quantized, and perhaps even thrust.

- ## Spectroscopic analysis

Spectroscopy uses radiation to investigate chemical compounds. It's just one of a whole range of techniques now available to the modern chemist – techniques which can determine the structure and composition of various substances. Other analytical tools include **mass spectrometry** (used to establish the identity and structure of unknown molecules) and **X-ray crystallography** (used to ascertain a molecule's 3-dimensional shape). Ironically, a chemist wishing to investigate a particular compound will often seek to isolate the same, often using **chromatography**. Isolating such compounds actually hints at the corrupting influence of its surroundings, and the fact that spectral patterns are likely to be affected by the same. Because of this, purifying and isolating compounds makes a detailed understanding of telepathy much more difficult.

As mass spectrometry provides only a limited amount of information about unknown compounds it is often supplemented with spectroscopic forms of analysis, such as nuclear magnetic resonance imaging (NMR), otherwise known as **magnetic resonance imaging** (MRI). NMR and MRI both exploit the behaviour of carbon and hydrogen atoms (the body's foremost structural components) when placed in a strong magnetic field. With this technique, scientists can build-up a detailed picture of large organic molecules possessing pronounced carbon backbones. Infrared spectroscopy can then assist in identifying all the functional groups bound to those carbon structures, through an examination of all the various bonds.

Additionally, one might use X-ray crystallography to determine the 3-dimensional structure of these molecules, as famously happened in the 1950s when British X-ray crystallographer Rosalind Franklin (1920-58) produced X-ray diffraction images, which suggested that DNA might be helical. American geneticist James Watson (1928-) and English molecular biologist Francis Crick (1916-2004) showed that DNA was, in fact, the iconic double-helix shape. And this is just about as advanced as modern chemistry professes to be – deconstructing large organic molecules so as to understand their shape, structure, composition and the nature of their bonds.

Whilst scientifically profound, this is still not the whole picture. For remotely-induced effects to be a reality requires more than a complicated arrangement of atoms and molecules, bathed in so many cations and anions. It requires a unique set of variables impacting on a given electron, sufficient to make each jump exclusive to the individual. Mastery of that chemistry and physics was achieved, relatively easily, in the first half of the 20th century. All that remained, however, was mankind's mastery of itself.

PART II

Induced Effects

Chapters 4-6

Ch4:   **Remote manipulation**

•   Unlocking the destructive

The human brain's latent telepathic potential was unlocked sometime between Niels Bohr's 1913 treatise (On the Constitution of Atoms and Molecules) and the Shelter Island Conference of 1947. One is able to argue in favour of such an early date because the chemistry which permits such phenomena to exist is fundamental, rather than specialized. That on a sliding-scale between base-chemistry and complex biological adaptation, we're talking base-chemistry – chemistry so basic that just one-step up from chemical bonding we encounter a whole new paradigm of material awaiting scientific analysis. One way to examine this relationship between base-chemistry and neurophysiology is to contrast the development of the eye with that of the brain.

The brain appears to have evolved before the eye, with the eye borrowing heavily from the brain, i.e. both have the capacity to absorb energy and convey it, in the manner of a signal, to other parts of the body. Where the brain and eye differ is in respect of the spectra themselves. The brain utilizes unique spectra, whereas the eye draws upon a common pattern of absorption. In other words, we all see the same blue sky – save and except those with defective colour-vision. This is explicable due to the fact that personal biomagnetism is too weak to affect absorption in the visible. However, such biomagnetism is strong enough to generate absorption bands or lines at much longer wavelengths. In this way we all see the same arrangement of colours, but think about them as distinct individuals.

Back in 1887 Heinrich Hertz became the first person to send and receive radio waves, in a simple experiment which did little more than induce a spark across a small gap. Shortly afterwards it became possible to create the neurological equivalent of that spark, by triggering the **action potential** of a given neuron, sufficient to bring about the firing of an associated synapse. The absorption of that initial radio signal would have caused sodium ion channels to open in the plasma membrane of the said neuron, flooding the inside of the cell with positive charge. That sudden **depolarization**, as it's called, may have visibly affected the subject, they being the very first person to be remotely manipulated.

The expression "telepathically-induced effects" includes remote manipulation, mental telepathy and the superimposition of persons.

All of which are produced using man-made devices – which, for the sake of convenience, we'll term radiofrequency technologies. Therefore, an "induced effect" is, by definition, man-made, and consciously applied in order to produce a given result. Remote manipulation is simply the most rudimentary of all the phenomena discussed in this book. More sophisticated is mental telepathy, and beyond that lies the superimposition of persons. British novelist Virginia Woolf (1882-1941) was contemporaneous with this potential being realized, describing herself as "a sensitive plate exposed to invisible rays" in one of her many works.[15]

• Sensitive plates and the photoelectric effect

Einstein's paper on the photoelectric effect (first published in 1905) showed that electrons could be dislodged from a metal plate, provided the plate was exposed to radiation of a particular wavelength. Likewise, Hertz had established many years earlier that an electrical current could be induced in an aerial or antennae through exposure to radiation of a specific frequency. And, it was also known that some forms of radiation would pass unhindered through these metal objects. Leaving aside the fact that this process of **electromagnetic induction** is far more sophisticated in living organisms, we can nevertheless discern a fundamental truth in all these observations – namely, that an electron may or may not absorb a photon; but when it does an electrical current is possible (most conspicuously in aerials and antennae, due to the nuclei sharing electrons).

Who knows, perhaps those dastardly hydroxyl free radicals help to maintain a useful differential within living systems. What we can say, however, is that photons comprise oscillating electric and magnetic fields, which are sometimes so strong they're capable of dislodging electrons from their parent atoms. At other times those fields contain just enough energy to be absorbed, whereupon they influence the movement of the electrons, but with the electrons still firmly bound to their nuclei. Or, alternatively, those fields may be weak, resulting in the radiation passing straight through the atoms, molecules and bodily tissues. An electron is therefore either: 1) wholly unaffected by incident radiation, 2) affected as regards its distance from a given nucleus, or 3) removed from the parent nuclei altogether under the influence of **ionizing radiation**.

When an electron moves it influences the movement of other electrons, whilst all the while being under the influence of those fields produced by other electrons. Fields therefore influence electron behaviour – ranging from wholesale ionization at one extreme, through molecular bond vibration, to the generation of electrical currents, not to mention electronic effects affecting just one molecule, and hence a given individual. This range of effects is at odds with the presumption that radiation acts in a uniform manner, with the reality being far more complex than you would have initially imagined. The human brain reflects this diversity, having evolved to avoid the kind of energies which are likely to be absorbed from external sources. Moreover, it's encased within the cranium, which is bone, and therefore resistant to the destructive effects of ionizing wavelengths.

Personal biomagnetism therefore mediates between the myriad of electrons found within a given human brain – electrons which affect principally themselves. Any absorbed energy isn't simply spat-out, or emitted, any more than the eye emits what it has absorbed. Instead, that energy is re-distributed within the neuron, triggering various chemical processes; ones with the capacity to create nerve impulses. This interpretation of the mind paints the neuron as being, quite literally, the mind's eye – with **Homo sapiens** having evolved to be a casual observer of its own thinking, due to the human brain absorbing and emitting within itself. This makes the human mind a semi-isolated phenomena, whose success owes much to that sense of isolation.

•   Transduction of energy

How the nervous system responds to stimulation is, perhaps, *the* central pillar of neuroscience. Broadly-speaking, **receptor cells** (cells which are sensitive to touch, taste, smell, sound and light) convert the energy associated with such stimuli into nerve impulses. To do this a certain threshold must be exceeded, called the **local potential** (or receptor potential). The human eye, for example, consists of a lens, which focuses light onto a person's retina. The retina contains **photoreceptor cells** (specifically rods and cones), which together are sensitive to both light-intensity and colour. Although the mammalian eye is highly evolved, the chemistry driving binocular vision is still, fundamentally-speaking, the same chemistry which evolution originally seized upon and developed. It is these most elementary chemical processes which facilitate the remote manipulation of individuals.

The human eye possesses photosensitive pigments, all of which comprise a protein molecule bound to a **retinal molecule** (11-cis-retinal). The protein molecule determines which wavelength of light is absorbed, and hence which colour is perceived. Once absorbed, the associated retinal molecule changes its shape (becoming 11-trans-retinal), in a process known as **photoisomerization**. In essence, the light absorbed by one of the molecules causes the other to alter its shape, triggering a chain of events which leads to a nerve impulse. And that, as they say, is that. The whole of colour vision hanging on an alteration to the shape of one particular molecule. The fact that **hyperpolarization** then follows, in the case of colour vision, rather than depolarization, is somewhat academic. What's important is that light-energy is able to initiate a whole series of intracellular reactions relatively easily.

Figure 8: Second-messenger system

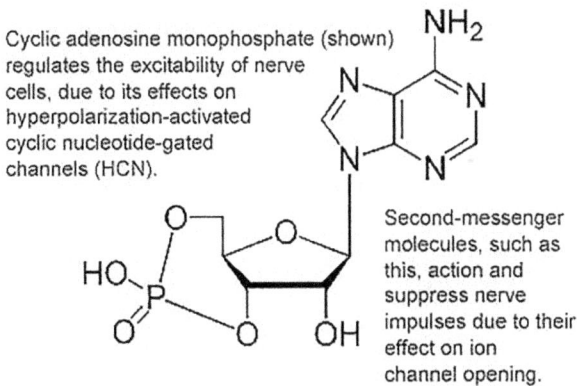

Cyclic adenosine monophosphate (shown) regulates the excitability of nerve cells, due to its effects on hyperpolarization-activated cyclic nucleotide-gated channels (HCN).

Second-messenger molecules, such as this, action and suppress nerve impulses due to their effect on ion channel opening.

And this relative ease explains why the human brain is configured to avoid the corrupting influence of incident radiation – primarily by having an absorption profile unique to the individual. Simple **isomerism** is able to trigger nerve impulses; with unique spectra avoiding the same. If a molecule in a neuron is made to change its shape as a result of incident radiation – that molecule having the same number and type of atoms, but with the atoms adopting a different arrangement – it has the potential to trigger a whole series of chemical processes within the cell. One of the processes it might activate is known as the **second messenger system**. It is the second messenger system which catalyses enzymatic reactions within the neuron, sufficient to activate sodium ion channels.

Second messenger molecules, such as cyclic adenosine monophosphate (see Fig. 8) directly influence ion channel opening, making this kind of intracellular reaction an obvious target for those wishing to remotely manipulate individuals. Control over

47

these neuronal activities provides for third-party control over all the processes which sustain life, action bodily movements and which initiate cognitive functions. Therefore, this neuro-cognitive potential is arguably mankind's most fearsome capability. By 1947 those giants of western political and military intrigue were systematically burying this truth, amid burgeoning Cold War politics and the distraction of a growing nuclear arms race. Arguably, the Western Allies provided for the Soviets successfully testing an atomic bomb amid the growing debate about QED, precisely because it acted as a useful diversion. This well-intentioned act of carefully-constructed carelessness has, in fact, served **The West** very well indeed, what with the cream of Russian science fixated on the atomic nucleus – like Heisenberg and the fascist Reich before them. Herein lies the roots of a much-needed unity; one which has usefully bound the Western Allies ever since, and in a manner transcending fractious mainstream debate. A unity which is only now facing extinction, with the rise of ultranationalist tendencies in Britain and America.

- **Mass manipulation**

Once a nerve impulse has been achieved, **sodium inactivation** returns the neuron to its original resting state. So that control over sodium ion channel opening *and* sodium inactivation provides for complete control over all the various **electrical potentials** associated with a given neuron (be it the resting, local, threshold or action potential). With this technical mastery comes absolute control over a neuron's electrical state – and with it the potential domination of a given individual. Clearly some people would have been more productively imposed upon than others in the early postwar era, with the remote manipulation of individuals confined to a select few. But let's not be drawn into a premature debate about the fate of the wider body-politic – but rather, examine the issue of remote manipulation in greater detail.

Mankind's ability to transmit a particular wavelength, one which can only be absorbed by a specific person's brain tissue, is due to the somewhat varied nature of electromagnetic induction. In those areas of the electromagnetic spectrum, way beyond wholesale ionization, indiscriminate bond vibration and telecommunications, we enter a region of individual electronic effects. And these effects needn't be dramatic – they might be nothing more than a slight re-configuration of a single molecule in a given person's brain. Even so, if the enzyme action which results is effective in opening sodium ion channels, then the inside of the plasma membrane will depolarize and a nerve impulse will be triggered (all assuming, of course, that the action potential has been achieved).

Assuming that the action potential has been reached the nerve impulse will travel from the cell body down the **axon** to the synapse. At the synapse the signal must cross the gap between contiguous neurons in order to stimulate the nerve cell on the other side. **Neurotransmission** (see Fig. 9), which is the term given to the electrochemical bridging of this gap, is actually more complicated than the production of the original signal – with chemicals (called **neurotransmitters**) actively facilitating or inhibiting the passage of this signal. The most prevalent neurotransmitters in the brain are amino acid transmitters – which have the capacity to both excite adjacent neurons (e.g. glutamate, aspartate, etc) or inhibit their action (e.g. glycine, gamma-amino butyric acid, etc).

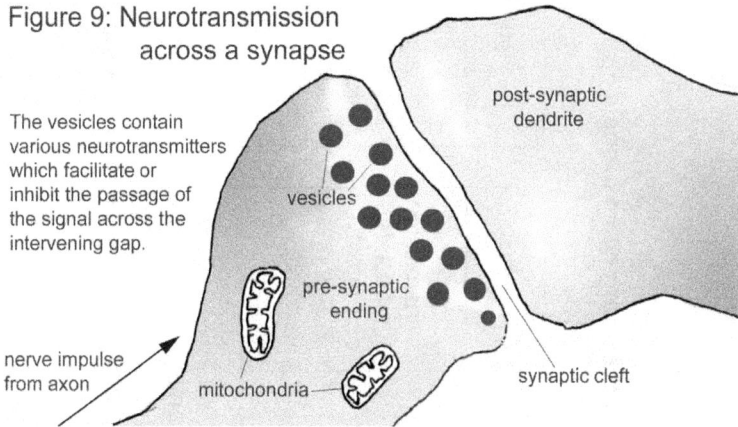

Figure 9: Neurotransmission across a synapse

The vesicles contain various neurotransmitters which facilitate or inhibit the passage of the signal across the intervening gap.

post-synaptic dendrite

vesicles

pre-synaptic ending

nerve impulse from axon

mitochondria

synaptic cleft

Pharmacologists, developing new drugs, often target synapses in order to replicate the behaviour of neurotransmitters – and many illicit drugs, such as heroin, cocaine and ecstasy, directly influence neurotransmission in ways which are likely to lead to chronic dependency. Obviously, manipulation *en masse* leaves the body-politic caught between potentially pernicious forms of interference on the one hand and the pharmaceutical industry's desperate attempts at remedying the same on the other. However, for the wider population to be this vulnerable requires more than convincing abstract models and persuasive scientific thinking, it requires tried and tested hardware and sound engineering.

Figure 10: Transmitting and receiving radio signals

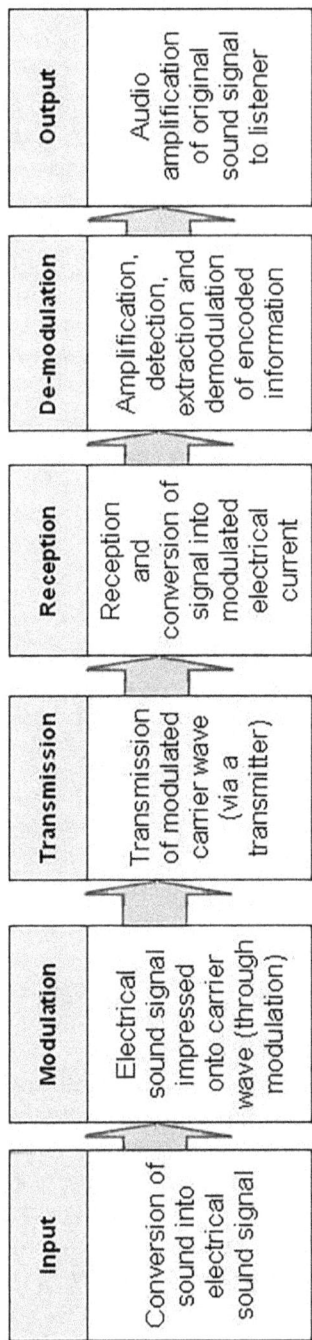

Input	Modulation	Transmission	Reception	De-modulation	Output
Conversion of sound into electrical sound signal	Electrical sound signal impressed onto carrier wave (through modulation)	Transmission of modulated carrier wave (via a transmitter)	Reception and conversion of signal into modulated electrical current	Amplification, detection, extraction and demodulation of encoded information	Audio amplification of original sound signal to listener

The key to transmitting meaningful sound signals to receiving aerials lies in the modulation of specific carrier waves - which was achieved originally through amplitude modulation, prior to frequency modulation being developed in the inter-war years.

- ### Transmitters and transmittance

Building on Hertz's very first radio transmission, Croatian born engineer Nikola Tesla (1856-1943) and the Italian born inventor Guglielmo Marconi (1874-1937) pushed radio technology to its absolute limits. First came Tesla's much vaunted "wireless telegraphy"; followed shortly after by Marconi's first transatlantic radio transmission, which took place in 1901. Marconi's radio signal (broadcast from Cornwall, England) was successfully picked-up in Newfoundland, a distance of no less than 2,232 miles (3,500 km). Capitalizing on his successes, Marconi subsequently helped the British Government to establish an entire worldwide radio network, which formed the basis of the British Broadcasting Corporation's Empire Service, later known as the BBC World Service.

For radio signals to be broadcast and received the following was needed: 1) a **transmitter**, with a height proportional to the frequencies being transmitted; 2) a receiving aerial, which would experience a small induced current when a signal was received; 3) a frequency selector, so that the right **carrier wave** could be chosen; and 4) an electronic circuit, to extract the relevant signal and amplify it through a speaker. Initially, in order to extract the right signal, a device containing iron-fillings was used; but this was subsequently replaced with **valves** (1904) and **transistors** (1947), which could extract, rectify and amplify signals far more effectively. So much so, that by 1904 radio technology was firmly established, becoming an integral part of daily life by the end of the decade.

British physicist, John Joseph Thompson had discovered that electrons, emitted by a negatively-charged cathode, were drawn towards a positively-charged anode (in the manner of an electric current bridging the intervening space). This discovery paved the way for the electron tube, or valve, which could greatly amplify a signal. This valve technology enabled the BBC's World Service to broadcast to all four corners of the globe, and it did so in an age well before satellite communications, using only ground-based transmitters. These transmitters, sited in British-held territories, used oscillating currents in their antennas in order to broadcast at specific frequencies (see Fig. 10). Clearly, applied physics and chemistry were beginning to utterly transform the character, shape and substance of everyday life – and, it seems, beyond everyone's wildest expectations.

Cathode rays and valve technology had the effect of amplifying science's understanding of the atom – and it's no coincidence that television was invented in 1926, just as the Copenhagen interpretation was gaining support. TV drew upon everything which had been learned as regards electricity, radio technology, cathode rays and the atom, in order to project moving images and associated sound into peoples homes. Because radio (like other forms of electromagnetic radiation) has the potential to be reflected off objects, radio detection and ranging, otherwise known as **radar**, was developed shortly afterwards. Radar rapidly becoming a vital component of air traffic control, navigation and defence.

The remote manipulation of a given individual would have presented far fewer difficulties than receiving a signal from a specific person – which would have required detection, extraction, rectification and amplification (not to mention the manufacture and maintenance of all the equipment, together with its periodic servicing). However, these are problems of scale, rather than intractable difficulties; and the free availability of valve technology would have made amplifying the received signal that much easier. It was therefore within the capacity of a scientific elite to perfect such technology, albeit on a modest scale, prior to the postwar era.

The expertise and technical know-how needed to generate radiofrequency transmissions in this very precise and predictable manner was readily available by 1913, and very highly developed by 1947. Therefore, by the interwar years, remote manipulation, mental telepathy and the superimposition of persons were all technically possible. It is an exciting and perturbing irony that as electromagnetic induction becomes ever more person-specific in the longest conceivable wavelengths, every other object becomes wholly transparent to the radiation in question. This means that the photon which is capable of affecting your thoughts, actions and behaviour can pass unhindered through every single intervening object. Whenever such a signal, photon, transverse wave, elementary quantum of action, however you wish to describe it, passes through an object in this manner, it is termed **transmittance**. Such transmittance makes all attempts at hiding from induced effects impracticable.

- Thanks to the ionosphere

The ionosphere is that part of the upper atmosphere which contains appreciable amounts of ions and free electrons. Electric and magnetic fields produced by the sun are so powerful they tear electrons from their parent atoms leaving an ionized layer in the Earth's upper atmosphere. This ionized layer is known to facilitate or hinder telecommunications across the globe due to both daily and seasonal variations in its reflective properties. The world of telecommunications is therefore awash with terms such as ionospheric forecast, ionospheric storm, ionospheric disturbance, etc, all because of phenomena which could, in theory, hamper radio communications.

Figure 11: Effect of ionosphere on radio waves

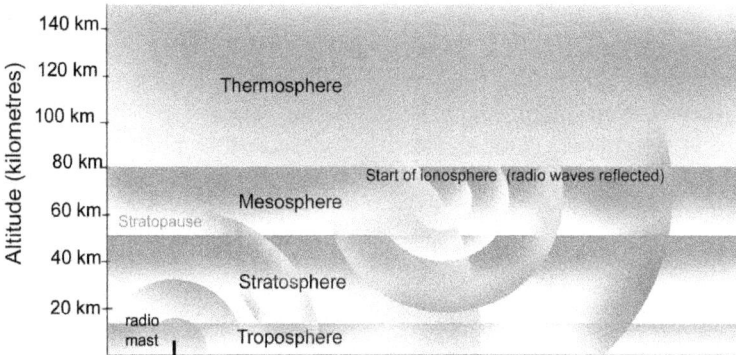

Generally speaking the ionosphere (see Fig. 11) facilitates radio communications because electromagnetic waves longer than 10 metres are normally reflected back to earth. Therefore, if the signals which are absorbed and emitted by the human brain have wavelengths in excess of 10 metres, then ground-based transceivers would be sufficient, both technically and strategically. Any shorter than this and geostationary satellites would provide a better platform from which to remotely influence individuals. The evidence suggests that person-specific absorption profiles comprise wavelengths much longer than those used in regular radio communications. Consequently, the ionosphere acts like a perfect mirror to such radiation, with the radiation in question being highly penetrative as regards ground-based objects.

By 1947 the Western Powers had perfected the sending and receiving of radio signals to and from specific individuals. These signals would reflect neatly off the upper atmosphere or propagate as **groundwaves** by following the curvature of the earth. Invariably their target was the central and peripheral nervous systems of named individuals. The peripheral nervous system includes the **autonomic nervous system**, which comprises two sets of nerves, i.e. the sympathetic and parasympathetic. Working in concert, these sympathetic and parasympathetic nerves regulate the body's

53

panoply of organs, glands and secretions, sufficient to sustain life. At the opposite end of the neurological spectrum, so to speak, lies the central nervous system's principal organ, the brain – which is tasked with the maintenance of both individuality and self.

In many ways the autonomic nervous system is the obvious target for remote manipulation, as it provides for doctors, medics and physicians directly intervening in the processes underlying the general maintenance of life. But this is very much the tip of the iceberg as regards telepathically-induced effects. Contrary to what purely inductive forms of reasoning might suggest, it's become possible to subject multiple individuals to a plurality of simultaneous effects, whilst integrating their every conscious thought; and with some cognitively subordinated to others. How such individuals might subsequently interrelate, via such technologies, is the subject of the next chapter.

Ch5: **Mental telepathy**

- Modulation and neural switching

A carrier wave is a radio wave, produced by a transmitter, onto which information is then impressed using, for example, **amplitude modulation** (AM) or **frequency modulation** (FM). **Modulation** impresses information onto the wave by altering its fundamental characteristics. In the case of AM radio, the height of the crest alters; but with FM radio, the wavelength and frequency simultaneously fluctuate. Modulation, in the case of mental telepathy, would amount to little more than triggering the action potential of relevant neurons – possibly using a derivative of FM, in order to stimulate a number of **ion channels** within a given bandwidth. Far simpler, one suspects, than "double-sideband amplitude modulation", "phase modulation" or "pulse amplitude modulation", to name but three, of all the various forms of modulation associated with conventional telecommunications.

But if mental telepathy amounts to a simple all-or-nothing firing of specific neurons, doesn't that make it a **binary system**? The binary system utilizes just two digits (0 and 1) to create the equivalent of a switch, i.e. 0 (*switched off*) and 1 (*switched on*). A binary string comprises several of these switches, which are either in the on or off position. A binary string therefore consists of so many bits, each of which is either a zero or a one. For example, a binary string comprising eight bits, i.e. a byte (by eight), can represent 256 possible values, simply by arranging the zeros and ones in every conceivable way. Moreover, this all-or-nothing digital approach creates crisp, noise-free results, because signal distortion rarely affects the result.

And so it is with the human brain, the action potential of each and every neuron creates a crisp, noise-free result, unaffected by extraneous chemistry or superfluous electrical activity. Each neuron is therefore a well-conceived switch in the mother of all **digital electronic circuits**. And, as American mathematician Claude Shannon (1916-2001) established in the 1950s, it's possible to replicate complex neurology, albeit in a very scaled-down form, by using the kinds of binary algebra found in modern computers. What's missing from these relatively crude attempts at artificial intelligence is the ability of the replica mind to both absorb and emit within itself. Simulate this, and it may be possible to train people using virtual subjects; the computer becoming a casual observer of itself and its surroundings.

British mathematician Alan Turing (1912-54) famously conceived of a test of this kind of artificial intelligence, in which both a computer and a person are interrogated by a questioner, with all three isolated from one another. Both the computer and the person deliver their answers, and it's then up to the questioner to decide which is the computer. If the person interrogating can't determine which is the computer, the computer is said to have passed this so-called "Turing test". A more advanced version of this test would involve connecting man and machine telepathically, i.e. taking the thoughts of the machine and impressing that information onto the carrier waves of the person, and vice versa.

This process of switching (to borrow a term commonly used in telecommunications to denote the process of connecting two users) would involve entire absorption profiles, rather than single isolated carrier waves. This ability to convert signals from one person to another, what we might term **neurological switching**, takes us beyond the mere remote manipulation of individuals, and into the realms of mental telepathy and the superimposition of persons. As a convenience we can say that such effects are applied bilaterally or multilaterally, with **bilateral effects** arising when a single person is remotely manipulated, whereas **multilateral effects** arise when several people are united by simultaneous involvement. Mental telepathy necessarily involves multilateral effects; and we could, for the sake of convenience, term this common meeting-ground **neurospace**. This psychological equivalent of personal space obviously begs certain questions regarding how much autonomy, privacy and distance individuals should be afforded.

Whatever those concerns, provided those responsible for the imposition of such phenomena are suitably equipped with a whole series of wavenumbers, ground-based transceivers, and a sound ionospheric forecast, there's no reason why mental telepathy shouldn't become front page news in the 21st century. With its global reach, tried and tested chemistry and the ability to penetrate any structure, such phenomena could easily imbue the pilot, doctor, aid worker, peacekeeper or diplomat with additional resources in times of crisis. Others, however, might feel cornered by the experience – and so we now turn to the issue of cognitive psychology, and the subject's state of mind.

• **States of mind**

A person subject to such effects often responds initially with **fatalism**, leading to one of three possible outcomes: 1) learned helplessness, i.e. failure to take action, due to a sense of not being in control; 2) agentic subordination, i.e. the individual's independence, autonomy and conscience are suppressed; or, alternatively, 3) a **libertarian** mindset, i.e. the person overcomes their initial fatalism, and any inhibitions, and resolves to be as autonomous as possible.[16] The first two mindsets constitute continued fatalism – fatalists subscribing to the doctrine of **predestination**, which means they perceive themselves as being utterly powerless against fate, and that their destinies as individuals are predetermined. Learned helplessness and agentic conformity reflect such fatalism.

The third group, libertarians, believe wholeheartedly in free-will and that people should have complete freedom of thought, conscience and behaviour (and shouldn't automatically be subject to the authority of others). Most people exhibit characteristics consistent with all three states of mind. For example, a libertarian might become an agent of others, of their own volition, only to end-up feeling powerless. Or, a person might be principally agentic and submissive, but respond autonomously under the influence of drink or stress. It's also fair to say that libertarianism doesn't necessarily make you a better person, it's simply perceived as being the more desirable frame of mind.

Figure 12: States of mind

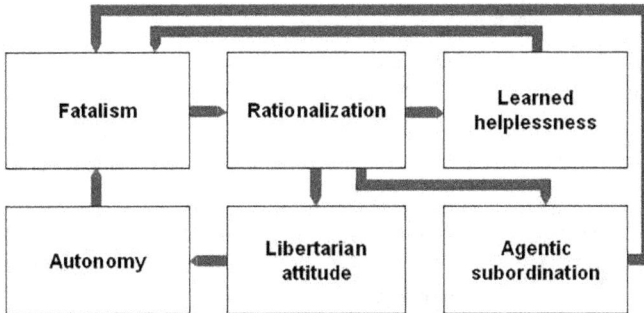

| Fatalism | Rationalization | Learned helplessness |

| Autonomy | Libertarian attitude | Agentic subordination |

All persons within the mainstream political environment are locked into this potential cycle - much depends on the forces bearing-down on the individual. Autonomy is no guarantee that those forces won't turn debilitating and oppressive.

The powers behind telepathically-induced effects might force the individual into a somewhat fatalistic outlook (initially, at least), whereupon the person is forced to re-orientate themselves, with or without assistance from others. Such is the complex nature of human psychology that to pass the so-called "Turing test" a computer might have to appear initially fatalistic. One instinctively views the alternative libertarian attitude as being somewhat defiant – combining, as it does, an uncompromising defence of the **self**, with an inflexible call to freedom and support for autonomous self-expression. The West has traditionally been tolerant of such calls for personal freedom, and even provided for a certain amount of rebelliousness on the issue.

As for the political extremes of both Left and Right, they would have sown the seeds of fatalism *en masse*, making predestination the official religion. Totalitarianism forces the individual to choose between agentic subordination and learned helplessness (see Fig. 12); and such is the intensity of the effects spoken of in this book that those states of mind would have manifested themselves in their most extreme forms. Independence, autonomy and conscience would have been replaced with a total disregard for the most vulnerable in society, with the worst-treated getting a very raw deal indeed.

• **Determinism versus free will**

Determinism is the doctrine or belief that everything (including human behaviour itself) is caused by something and that there is no free will. If predestination is a feeling of powerlessness regarding oneself, then **determinism** expands upon that to include all things. Belief in determinism may arise as a consequence of exposure to controversial ideas (for example, those related to genetic and biological determinism) or through living in a state which is entirely repressive. Fatalism reflects this conscious or subconscious belief in predestination and determinism, and contrasts with **free will**, which is defined as the ability to act and make choices as a free and autonomous person, and not as a result of compulsion or force.

Libertarians believe in free will, even if it seems like a distant goal, rather than a fact of life. And the true libertarian will seek to maximize free will for all, rather than using the same to predetermine the lives of others. It is within this libertarian ideal – of autonomous self-determination – that mental telepathy and associated effects should be applied. Telepathy should be symptomatic of personal freedom, rather than evidence of more questionable politics. But subscribing to free will is one thing, actually arriving at the ideal political and social framework is quite another. For example, are the factors governing our politics and behaviour ultimately the result of determinism, whatever our stated values and beliefs?

One day we may succeed in producing a computer which absorbs and emits information within itself and about itself, and which therefore replicates human consciousness. At present they rely heavily on external information conveyed to them via the keyboard or mouse. But such a computer would still be vulnerable to external interference, and so it is with humans. However much conscious self-awareness and intelligence they can muster, people remain vulnerable to external interference, some of it destructive. This is why the political environment, so heavily fought over in the last one hundred years, is so vitally important.

The political framework associated with the democratic free-world comprises a more-or-less immutable constitutional structure, within which day-to-day politics arises. That uncompromising, inflexible and defiant fight to the death with the forces of the radical right, which took place between 1939 and 1945, was about the preservation of these constitutional arrangements; and ultimately about adding to them in an equally immutable fashion. So much so, that fractious party politics could be increasingly indulged, fostered and supported within the postwar environment, and in spite of the overarching arrangements growing and becoming ever more inflexible.

And this appears to be the key to libertarianism and free-will – having a sound constitution; one which supports freedom of thought, conscience and behaviour, and the diverse politics that commitment invariably gives rise to. By the start of the Cold War the Western Powers were busy adding to the existing constitutional apparatus, albeit in a highly clandestine way. This whole new socioeconomic apparatus would, in theory, provide for the legitimate use of radiofrequency technologies, without imposing egregiously on mainstream politics. Conventional politics, particularly in America, began to take on a highly-charged tone, as this unchallengeable construct materialized in secret. Those not party to such secrets would, of course, be left wondering as to the cause of events.

• Radiofrequency attribution

Attribution theory suggests that people formulate theories about why things happen using an *a posteriori* approach (arguing from observed effects to unknown causes). Attribution theory, which was devised by Austrian psychologist Fritz Heider (1896-1988), recognizes the following types of attribution: 1) **dispositional**

attribution, in which personality is perceived as being the principal cause of an action or event; 2) **situational attribution**, in which circumstances are seen as being the principal cause of an event or occurrence; 3) **actor-observer bias**, in which negative behaviour on the part of others is attributed to their character, but similar behaviour in oneself is blamed on the situation; and, 4) **fundamental attribution error**, in which too much blame is apportioned to another's personality, and too little significance given to their circumstances.

In many respects the determinism versus free will debate is an attributional one, in-so-much as free will implies there is no underlying cause to a given action or event, just individual choice. In reality, our capacity to make choices is always matched by a given situation's capacity to impose on the choices we make. One could argue that the constitution exists to support, in principal, complete freedom – but also those instruments which unavoidably impinge upon the same, such as the judiciary. The genius of American science was to marry radiofrequency effects to the constitutional apparatus, such that the only thing constraining free will was, in theory, mainstream politics and the law.

Because the constitution provides for freedom and democracy, and because these catalyze both politics and the law, then the electorate is nominally imposing upon its own free will through the latter. My point is this, by becoming a part of the constitutional framework, mental telepathy wouldn't, in theory, impose on free will – but neither would politics and the law impose on mental telepathy. From its inception, the intention was to afford people free will, constrained only by the demands of their own chosen politics – a policy which provided for two world wars and extrasensory impositions. In the case of Nazi Germany, it was the disposition of its Fuhrer and the character of its people which plunged Europe into war – but they, of course, would argue it was their wider circumstance.

Admittedly, fundamental attribution errors do arise whenever too much importance is given to another person's character, and not enough weight to the forces at work within that person, or indeed, their wider circumstance. Today, it's the medical profession or the law which is most likely to make fundamental attribution errors, by giving too little credence to those **exogenous factors** spoken of throughout this book. Much depends on the way in which neuro-cognitive effects are applied. But, given the natural-born tendency of most people to blame the individual, it seems only fair that mental telepathy and associated phenomena are made public knowledge – if only so we can put them on a sound constitutional footing and protect the individual.

- **Memories, ideas and intentions**

The total set of memories, ideas and intentions which a person holds about himself or herself is termed their **self-schema**: this aspect of the person's personality being the principal target for those engaged in dispositional attribution. The Axis Power leadership, high-profile Soviets and many key-figures in the West have all successfully created for themselves very well-defined self-schemas. Self-schemas which draw heavily upon the selective memories, cherished ideas and brutal intentions of both themselves and their compatriots. Such was the forceful nature of those ideas, memories and intentions, during those long years of total war and ideological conflict, that remotely-induced effects would have done little more than supplement the same. Therefore, we can be quite certain that the individuals involved were true to their respective causes.

This isn't to suggest that high-profile figures weren't subject to radiological effects during the last war, only that their respective self-schemas were already heavily delineated by the individual's own chosen rhetoric, concrete ideas and given manifestos prior to the effects being applied. As those responsible for the effects were constitutionally in favour of affording such people a free hand, these particular world leaders would have been constrained only by the demands of their own and each other's politics, and by the political systems which spawned them. This was therefore an age in which the political Centre became thoroughly educated as to the hazards posed by man's true nature – and, the terrifying effects of knowledge, given the true nature of man.

The capacity to read minds, remotely manipulate physiology, transfer thoughts and superimpose upon individuals all materialized between 1913 and 1947. A decision was therefore taken early in the 20th century to eliminate the extremes of Left and Right, prior to the Western Powers divesting themselves of such secrets. Mental telepathy could only be safely assimilated in a world free of wholesale repression, and with the right constitutional apparatus conscientiously put in place. No one in possession of this neuro-cognitive capability doubted for a moment that totalitarianism would lead to mass conformity, learned helplessness and a fatalistic future – with the worst treated meeting with unspeakable cruelties.

This book postulates that the most appropriate constitutional structure is pro-democracy, supportive of free will and libertarian – in-so-much as it provides for all things. What that democratic structure contains is an executive, legislature and judiciary, whose actions serve to restrict complete freedom, ostensibly with the consent of the electorate. Moreover, telepathically-induced effects are applied as part of the wider constitutional framework – and in a manner which doesn't interfere with individual freedom. It is therefore left to mainstream political instruments to define the limits, with an overarching presence looking on. And, whilst this political climate may appear highly-charged, it is also extremely illuminating.

From this all-embracing psychometric test are drawn both mainstream leaders and future guardians of the constitutional apparatus. Apparatus which provides for all things, without being all things itself. One which won't frustrate free will – but might assist those with such a mandate, should it be pertinent to do so. An environment in which sufficient is known, and the right arguments promulgated, for its presence to be scarcely mentioned. And, where the history which is taught enriches our appreciation of past sacrifices, without ever detracting from the need for some urgent revisions.

At the heart of all this lies the inimitable self – the evolutionary endgame, as regards hitherto unheard of spectra. Your profile being a never to be repeated pattern of absorption and emission, and the chemical basis for your spirit – a spirit which is married, whether you like it or not, to the constitutional apparatus – one which provides for all things: not least of which, the testing of your own character.

- ## Innate characteristics

The term self has both a physical and psychological dimension. In medicine, for example, a distinction is made between the body's own tissue, called simply **self**, and foreign tissue, such as a transplanted organ, which is termed **not self**. So, even the human body helps to define what is the unquestionable self, albeit in a rather elementary manner linked to the person's own chemistry and genetics. Additionally, the person-specific absorption profile sets the person apart from all other humans by adding to that conscious sense of individuality and uniqueness. Radiofrequency technologies challenge this inherent isolation by fusing together disparate psychologies and thought processes.

Given that thoughts, feelings and physiologies arising in one person's mind may compete, on occasions, with those arising in the mind of another, the self equates with **innate characteristics**, i.e. those characteristics arising within a given individual, rather than via some external source. Those innate qualities do risk being overwhelmed from a young age if such technological assimilation isn't handled correctly. And, whilst the inability to express oneself due to tension, anxiety and self-consciousness may be barriers to free expression, this isn't necessarily a good enough excuse not to assimilate telepathy into our everyday lives at the earliest opportunity.

Viewed in **Freudian** terms the mind, **psyche** or self comprises an **id** (which is the unconscious source of primitive drives and desires), an **ego** (which provides for consciousness, memory and planning), and a **superego** (that part of the mind which acts as a conscience to the ego). Some might warn of conflict arising when a person is drawn unwillingly into multilateral involvement, citing the **frustration-aggression hypothesis** as evidence of what dissatisfaction can lead to. Modern psychologists, however, paint a far more complex picture – acknowledging that discontent may, in reality, lead to wholesale acquiescence, debilitating depression and psychiatric disorder. Invariably, though, the libertarian mindset is *the* most aggressive, especially when confronted by the forces of determinism and predestination.

Civilization, so fiercely contested in two world wars, consists of a coherent constitutional framework, one which is supportive of various systems – be they political, economic, social or agrarian. Heavily dependant upon its agricultural sector, modern civilization comprises so many urban centres, supported by a single integrated economy. Urban success or decay ultimately rests on the wider economy, and strengthening and expanding that economy makes socioeconomic sense. So, although extrasensory effects could lead to agentic conformity and learned helplessness, they might equally improve the efficiency of the **free world**, both at home and abroad (which would be anything but frustrating).

The state's monopoly on violence ensures that nothing has a greater capacity to inflict harm, rightly or wrongly, than the **state** itself. This violence is a reflection of the politics of the state, and, in a democracy, its people. But the constitution one envisages would transcend such violence, together with the people and politics which spawned it. This is the point at which we start to think of the constitutional apparatus as transcending national boundaries.

Free will applies as much to the personality of the state as it does to the disposition of its people, and so this apparatus would have to accommodate a world of difference. But, whilst doings so, it would be integrating their economies, strengthening their cities and helping states and individuals to cultivate their own chosen self-schemas.

Ch6: Superimposition of persons

- Enzymes and adsorption

The absorption of transmitted radiation, by a neuron, results in a whole series of chemical reactions; triggered, it would appear, by little more than isomerism. All such reactions have **activation energies** – in other words, the amount of energy which is required to make specific molecules take part in the reaction. Interestingly, the magnitude of the activation energy is closely related to the speed or rate of the reaction. The greater the activation energy, the slower the reaction. Consequently, extremely fast neurological reactions correspond with minimal amounts of energy. Therefore, as luck would have it, radiofrequencies trigger high-speed reactions in the brain using negligible amounts of energy – energy with the propensity to pass through every single intervening object.

Thus radiofrequencies, containing negligible amounts of energy, nevertheless deliver sufficient energy to drive cognitive processes. The principal reason why low activation energies correlate with the fastest of chemical reactions is down to organic **catalysts**, known as **enzymes**. Without enzymes we wouldn't be having this debate, as they speed-up reaction times enormously. Without them second-messenger processes might take years, decades, or even centuries to complete. They make normal neurological functioning possible, and provide for the kinds of phenomena spoken of throughout this book. Therefore, the ionosphere acts like a perfect mirror to transmitted radiation, which then triggers the swiftest of neurological reactions, having penetrated the thickest of walls.

Enzyme action oils the wheels of superimposition (i.e. the transmission of cognitive processes, including actions and behaviour) and **psychokinesis** (i.e. the conveying of specific forms of movement). The ability to replicate neurological processes at break-neck speed is obviously critical in those instances where near instantaneous reflexes are called-for. Superimposition is therefore, to all intent and purposes, the superimposition of the principal mechanism employed by enzymes; namely, **adsorption**. Adsorption temporarily binds two **reactants** onto the surface of an enzyme, creating a distortion and hence weakening of their covalent bonds. This weakening of those bonds lessens the amount of time and energy needed to bring the two reactants together.

It's important to note that enzymes are completely specific; that they're not altered by the reaction; and do no more than make the reaction faster. The superimposition of such reactions at the neuronal level, and the near instantaneous feedback which results, creates the impression of being in that person. Or, if you're the subject, the sensation of being superimposed upon. The danger, of course, is that the subject's sense of self becomes diminished, with that person's innate characteristics being compromised. People who experience conflicting thoughts, feelings and behaviour are said to be afflicted by **cognitive dissonance**. And, whilst holding an inconsistent attitude *is* possible, and often without distress, the superimposition of persons can nonetheless produce inner conflict due to an individual's behavioural or **conative dimension**.

This conative dimension pertains directly to the will of a person, including a person's tendency to act according to their own expressed attitudes, impulses and desires. The term conative, or conation, is defined as the person's own expressed will; as opposed to the will of a third party directing their actions. In the case of a person heavily superimposed upon, one would need to strip away third party affectations in order to arrive at the subject's own conative dimension.[17] In effect, excessive neurological interference undermines the subject's will, obstructs the conscientious crafting of their character and generates conflicting cognition in the process. To avoid fundamental attribution errors occurring, greater openness is called for.

- Hypothalamic confusion

Many texts work inwards towards individual neurons and their distinctive chemistry. This book, generally speaking, works outwards from such chemistry, as whole brains aren't superimposed onto one another, so much as their individual chemical reactions. Having studied those intracellular reactions in a suitable amount of detail we now turn to the structures which combine to form the brain. The human brain consists of a **forebrain**, **midbrain** and **hindbrain**, which together constitute the central processing unit or control centre for the entire nervous system (see Fig. 13). The central nervous system, as we've seen, consists of the brain and spinal cord, which extends out to the extremities and sensory organs via the peripheral nervous system.

The forebrain comprises the two cerebral hemispheres, simply termed the left and right hemisphere (together called the cerebrum), and the diencephalon (which contains the thalamus and hypothalamus). The **cerebrum** is the most advanced portion of the brain, being the seat of learning, sensory perception, language, emotion and cognitive processing. As for the midbrain and hindbrain, they control involuntary movements, posture and a range of unconscious activities. The hindbrain is the most noticeable feature after the forebrain, as it incorporates both the **pons** and **cerebellum**. Many of the centres controlling the autonomic nervous system's sympathetic and parasympathetic nerves are situated in the hindbrain, midbrain and hypothalamus; for example, those regulating heart-rate, respiration and hormones.

Figure 13: The Brain (general structure)

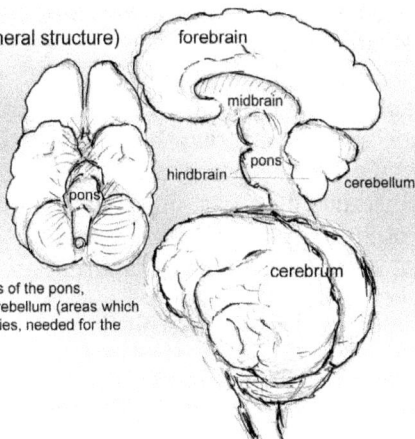

The forebrain comprises the cerebrum and diencephalon

The midbrain comprises the reticular formation, which regulates activity in the forebrain (triggering sleep if necessary).

And, the hindbrain consists of the pons, Medulla oblongata and cerebellum (areas which control unconscious activities, needed for the maintenance of life).

Men's brains tend to be larger than women's, but women's are said to occupy a greater proportion of their bodies. As for the **corpus callosum**, which is the tissue facilitating communication between the two cerebral hemispheres, it is relatively larger in females. Also, the interstitial nucleus of the anterior hypothalamus (**INAH3**), situated beneath the cerebrum (and which is responsible for male-typical sexual behaviour), is said to be two-and-a-half times larger in men, than women. Add to this the fact that men tend to lose their brain tissue earlier in the aging process, and overall, a greater proportion of the same – and the whole thing appears to add up to male and female brains.[18] This obviously has its corollary in those innate characteristics spoken of above, and which do so much to affect the personalities of those states professing sexual equality.

Down at the level of individual nerve cells it's generally considered to be the number of connecting **neural pathways** which is significant, rather than the actual number of neurons. However, the brain contains many more cells than it actually needs – which one could argue actively facilitates the re-absorption of energy, given that the addition and subtraction of all those magnetic effects will render some neurons more effective than others at communicating in this manner. What does seem clear is that those sexual differences between men's and women's brains may be reduced or amplified when bound together in some multilateral fashion. Much depends on the depth of that involvement and the characteristics of the people themselves.

Undertaken in an unrestrained manner the superimposition of male and female minds could give rise to problems of a sexual, physical or psychological nature. Multilateral telepathic involvement incorporating both sexes is practicable, provided that this meeting of minds involves principally the forebrain, and preferably no more than the cerebrum. Routinely marrying people together via the midbrain, hindbrain or hypothalamus could, in theory, affect those processes needed to sustain life and which pertain directly to sexual physiological differences. However, as many of the future guardians of this technology are likely to be women, we shouldn't be overly perturbed.

- Implanting an efference copy

The **cerebral hemispheres**, labeled left and right, as per the person's own left and right, have four lobes apiece (see Fig. 14). These lobes are termed the frontal, parietal, temporal and occipital lobes. The surface of each hemisphere is sometimes called "grey matter", or more technically, the **cerebral cortex**. The cerebral cortex is covered in ridges (gyri) and furrows (sulci), with the very deepest furrows (fissures) separating the various lobes. Beneath this grey matter lies "white matter", which consists of nerve fibres connecting all the various areas of the outer cortex, i.e. connecting those regions where the conscious and subconscious processing of information takes place. Thus, countless numbers of nerve cells, acting in the manner of so many switches, connect these vital areas (via those all-important neural pathways).

The largest of the lobes is the **frontal lobe**, which contains the motor cortex, premotor cortex and prefrontal cortex. These are the areas where movement is consciously initiated, behaviour repressed and speech formulated. Remote stimulation of the left hemisphere's motor cortex will result in movement on the right-hand side of the body, movement that the unknowing person will interpret as being entirely voluntary. Therefore, superimposition and psychokinesis frequently result in behaviour which is interpreted as intentional. These nerve impulses, which are conducted outwards (away from the brain), are termed efferent. Efferent signals, and the actions they produce, may leave an image of themselves imprinted on the person's nervous system, called an **efference copy**. This copy facilitates unconscious competency in the given action, ability or skill.

Figure 14: The left cerebral hemisphere (showing lobes)

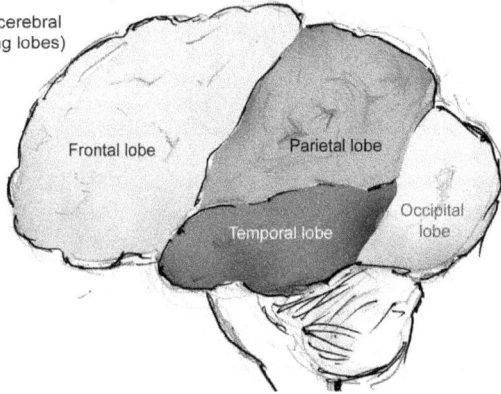

Frontal lobe

Parietal lobe

Temporal lobe

Occipital lobe

Unconscious competency enables routine tasks to be executed in a subconscious manner, perhaps whilst doing other things – the person having been initially incompetent, prior to achieving a certain amount of conscious competency, through concentration, prior to becoming unconsciously competent. The superimposition of motor skills leaves an efference copy in the subject, thereby facilitating such learning. Imparting skills – such as those requiring balance, memory, dexterity and coordination – could be made easier through the application of such effects. A whole range of people might one day be inculcated by others, using an efference copying approach.

Neurospace, as in the multilateral telepathic meeting-ground where interested people meet, is located primarily in the **prefrontal cortex**, situated at the very front of the frontal lobe. This is the region where people are most likely to be conscious of telepathy as a phenomena which they can engage with (and the area of the brain where the conscious transfer of thoughts takes place). The dark-side of this being the notorious prefrontal lobotomy, popularized throughout the 1930s, just as telepathically-induced effects were becoming more widely practiced. Such surgery may have been performed on patients distressed by such phenomena, thereby aggravating a potential injustice.

The prefrontal cortex is very much the seat of personality, conscious awareness, advance-planning and multilateral reasoning; in other words, the region of the brain which will ultimately determine whether neuro-cognitive effects persist or are consigned to history. This is because we're into the realms of hearts and minds – whereby issues are won or lost on the basis of people's wider perception. Winning over hearts and minds on these issues can't begin soon enough if such phenomena are to endure, becoming safely assimilated into mainstream society and culture.

• Sensory output and motor input

A given mind can be superimposed onto that of another, provided both absorption profiles are known. Those signals travelling from the subject's central and peripheral nervous system to the dominant person's mind are termed afferent; whilst those signals travelling towards the subject's brain and body are termed efferent. When this happens the dominant party (who we shall term the **superimposer**) will, in all probability, affect both the identity and thought processes of the subordinate individual. Normally such signals travel to and from the brain of a given individual within the confines of their own body. Superimposition has the effect of extending the path of both afferent and efferent signals to hundreds or even thousands of miles. And, as we've seen, this can be achieved at breakneck speed using negligible amounts of energy.

Afferent signals and **efferent signals** are conveyed to and from specific persons for the purposes of superimposition and psychokinesis. The above definition stretches the meaning of afferent to include signals arising in the mind of the subject, not simply those emanating from their sensory organs. Both types of signal would require neurological switching in order to match one absorption profile to the other. This combination of incoming afferent signals and outgoing efferent signals makes multilateral telepathic involvement and the superimposition of persons possible. Those afferent signals enable the superimposer to absorb precise information about the subject, which can then be responded to via efferent signals directed to particular areas of the subject's brain.

In order to determine which areas of the brain are associated with which particular tasks neuroscience uses a combination of magnetoencephalography (**MEG**), which detects small localized changes in the brain's magnetic field, together with functional magnetic resonance imaging (fMRI), which builds-up detailed images using molecules possessing large carbon backbones. In broad terms, the left hemisphere controls speech, language and calculations; and the right hemisphere spatial awareness, drawing and face recognition (see Fig. 15). Additionally, one hemisphere usually dominates – in right-handed people it's the left hemisphere. This is due to the **contralateral** nature of the human body, in which the left side of the brain controls the right-hand side of the body, and vice versa.

Figure 15: Specialization and the cerebral cortex

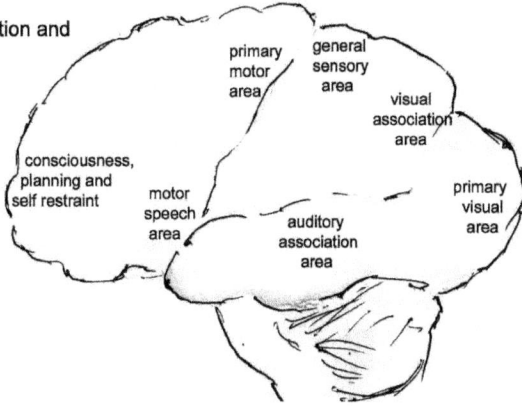

Because one hemisphere normally dominates, overall control rests with that hemisphere – including inhibitory impulses. So that a signal arising in the right hemisphere, telling the left-hand side of the body to move, can be inhibited by the dominant left hemisphere (in right-handed people). The impediment of inhibitory signals leads to **intermanual conflict**, more commonly known as alien hand syndrome. Alien hand syndrome could be remotely induced in one of two ways: 1) through the remote manipulation of chemical processes in the corpus callosum, preventing inhibitory signals reaching the side of the body initiating movement; and, 2) through the direct superimposition of motor impulses, sufficient to give rise to movement which the knowing person themselves interprets as being involuntary. Posterity will determine whether intermanual conflict arose in the recent past due to applied effects, but the evidence suggests that induced activity is more likely to be interpreted as voluntary (especially where the subject is unknowing, but knowingly interfered with).

The corpus callosum, which is the tissue facilitating communication between the two cerebral hemispheres, is said to be comparatively larger in women than men; which has led some to speculate that this facilitates communication between the two hemispheres, making women more conscious of their own and other's emotions when using language, applying radiofrequency effects, and analyzing the consequences of their own and other's behaviour, etc. Certainly, male domination of the politic tasked

71

with using radiofrequency technologies is unwise, and likely to result in the ill-balanced application of the same, due to some very male-typical psychology.

• **Sensing and perceiving**

Perception is perceived as a combination of bottom-up **modality** driven processes related directly to the five senses, combined with top-down conceptually driven ones, linked to the wider paradigm. That on a sliding scale between simple sensation and complex reasoning, perception lies somewhere in-between.[19] Speech is an example of something which can be sensed, reasoned with, and perceived. For example, one might hear an unknown language, but perceive only a lack of intelligibility. Hearing is a modality, and as such a category of sensation, e.g. touch, taste, sound, smell and vision. Telepathically-induced effects add to this list in an **extrasensory** way (quite literally, adding to the senses). But if such effects aren't to appear wholly incomprehensible we need to support them with a constructive paradigm.

How we perceive induced effects invariably affects the far more complex forms of reasoning which invariably follow. Fatalism, be it agentic conformity or learned helplessness, can arise as a consequence of what our five senses are telling us, but may also arise through extrasensory information derived through our minds. Extrasensory effects can therefore impact on behaviour directly, as with superimposition and psychokinesis, or indirectly, as a consequence of how such effects are perceived. **Gestalt psychology** treats behaviour and perception as one, and believes that these two aspects of psychology reflect the structure of the brain. Others might go further, seeing them as reflecting the wider political structure – particularly in a world of multilateral telepathic involvement, where totalitarianism serves to corrupt both.

Just like an out-dated paradigm, the five senses place strict limits on our perception. We can, for example, only discern clearly an area the size of a coin held at arms length and our hearing and smell are relatively poor. Whilst extrasensory information adds greatly to our modalities, actual perception continues to be frustrated by the present conceptual framework. Some of this is a reflection of political and military intrigue, but clearly a balance must be struck otherwise any strategic gains are rendered largely incomprehensible. Public awareness of these issues would provide for people's perceptions being responsibly shaped by those governing from as close to the political centre as is practicable. However, perceptiveness remains a powerful reward, and therefore an incentive to all those seeking to occupy the political middle-ground. By far the most dominant of all the senses, whatever our perception, is vision. So much so that not even extrasensory awareness can compete with this particular modality. Art, symbolism and design will always dominate civilization's outward appearance, with telepathy very much an adjunct to the same. In those instances where telepathically-induced effects do tangibly or discernibly affect mainstream society **radiofrequency attribution** will arise, though being a derivative of situational attribution one might find that both actor-observer bias and fundamental attribution error still dominate (even within this brave new world of potentially destructive interference). This brings us to **epistemology**, which deals with issues such as knowing, perceiving, inferring, wondering, proving, corroborating and being mistaken. Public knowledge would facilitate a sense of knowing, without which meaningful involvement becomes far more difficult.

· Paradigms, epistemology and power

Epistemology is that branch of philosophy which deals with the foundations, limits and reliability of knowledge. Epistemologists are often so ruthless in their description of proof that even orthodox science can seem provisional besides pure mathematics. Such ruthlessness, taken to an extreme, causes one to become a **sceptic**, i.e. one who denies that we can ever truly know something. And maybe we can't, if power is knowledge, and knowledge is withheld. The withholding of such knowledge might lead to widespread scepticism, and rightly so. But a conspiracy to do good, is still a conspiracy – right? With the sceptic being optimistic, rather than fatalistic. The real question is whether we know sufficient to provide for mainstream politics and law; so that provided we make the right choices the result is invariably judicious.

To avoid the wrong kind of scepticism we need to furnish society with a broad understanding of these issues. This general awareness should nonetheless be devoid of many of the finer technical details, in deference to the security of the free world. Official confirmation of such phenomena would nonetheless amount to a paradigm shift, with this book helping to define the limits of popular knowledge. Popper and Kuhn both agreed on the need for such periodic revisions, given the rather tentative nature of scientific knowledge. However, they differed significantly as regards the magnitude of such change. Popper perceived incremental revisions, arising within so-called normal science; whereas Kuhn anticipated something far more substantive and revolutionary, with the word "normal" having to be redefined.

Personal knowledge is equally tentative, with first-hand experience of neuro-cognitive effects causing many people to adopt and then abandon ideas, prior to them arriving at firm conclusions (conclusions which may amount to a full-scale revolutionary shift in their perception). This is in keeping with the ideas of Swiss psychologist Jean Piaget (1896-1980), who saw personal development as the gradual acquisition of logical abilities, many of them taught, which enable us to make increasing sense of the world. This process of personal growth can, nonetheless, be thwarted by political and military intrigue, frustrating those who would like to say with confidence that they truly "know" something. But knowledge, power and perceptiveness must concentrate in the right hands, and therefore be worked towards in some incremental historical fashion.[20]

Epistemologically-speaking we need only know enough to trust to our perceptions; more than that is a comparative luxury. Therefore, beyond the illusionary aspects of nature and the accepted limitations of the scientific method, we encounter the restrictions placed on us by global politics. In this context, the politics of power are complicated by the proffered constitutional structure transcending all national boundaries. This means that individual state's would be denied the luxury of knowing everything, but all would know sufficient to function in the manner of their choosing. This safely provides for states which are pro-democracy, supportive of free will and libertarian – and, crucially, all the things that such abject freedom periodically gives rise to.

Psychology has long pronounced on these issues, furbishing us with expressions such as **legitimate power**, expert power, informational power, reward power, coercive power, and so on.[21] This phraseology has taken on renewed importance due to the phenomena presently encroaching on all our lives. Swiss-born political philosopher Jean-Jacques Rousseau (1712-78) argued that the direct and continuous participation of all citizens in political life bound the state to a common good. Intruding on the physical processes underlying the citizen's every conscious thought, particularly at a young and impressionable age, risks frustrating that participation and replacing it with something more sinister and undemocratic. Properly applied such effects will actually strengthen the hand of the political middle-ground and provide for greater global freedom. However, systemic failings continue to threaten all that has been achieved – trapping numerous individuals in a difficult place, many of them children.

PART III

Self-determination

Chapters 7-9

Ch7:   **Absence of awareness**

- Pervasive developmental disorders

That capacity to read minds, remotely manipulate physiology, transfer thoughts and superimpose upon individuals may have materialized early in the 20th century, but it would be several decades before such phenomena would begin to impose routinely on the most vulnerable of all social groups – namely, the young. Basically, to be young and unknowing places you in a special category as regards vulnerability to induced effects. The young tend to live in-the-moment; whereas older people tend to engage in forward planning – whilst ever conscious of the past and whilst dealing with the present. Thus, the impulsiveness of youngsters is matched only by the capriciousness of those involving themselves remotely.

That impulsiveness and those caprices may lead to conflict. Sociologists have coined the term **hidden curriculum** to denote the informal codes that children are expected to conform to, but which aren't formally written down. Overuse of telepathically-induced effects risks bringing children into conflict with these codes, resulting in unnecessary criticism. Afferent signals arising in the child's mind are a powerful indication of the child's conative dimension and general psychology. Far more controversial, however, are efferent signals, imposed with the express intention of affecting the child's behaviour. The resultant efference copies, imprinted directly onto the child's nervous system, and the associated learned behaviours, may or may not be helpful to the child.

We've already seen how perception is a combination of bottom-up modality-driven processes related to the five senses, together with top-down conceptually-driven ones drawing heavily on what has been taught. But these effects aren't visited upon the child via the five senses and the broader conceptual framework is deceptive. This places the youngster at a potential disadvantage, with their broader development vulnerable to casual intrusion. Arbitrarily applied, such effects could, in theory, affect the young person's mental and emotional adaptation to new environments, be they social, academic or professional. Furthermore, fundamental attribution errors on the part of the psychiatric profession or law may stigmatize otherwise healthy individuals.

In addition to those conventional forms of power previously mentioned, e.g. expert power, reward power, coercive power, and so on, we might add **neuro-cognitive power**, to denote this potentially destructive presence. Neuro-cognitive power has the potential to subdue the **body politic** from an impressionable age, via routine involvement, leading to wholesale repression. Moreover, under conditions of excessive secrecy fundamental attribution errors may become endemic and widespread, resulting in far too much weight and importance being given to disposition, rather than circumstance. This might have the effect of frustrating the counselling process, leaving the person unresolved, conflicted and potentially suicidal.

It is particularly worrying to think that children are becoming increasingly caught-up in this clandestine use of extrasensory phenomena. Originally applied by a scientific elite to a strictly limited number of adults, one speculates that ever greater numbers, increasingly younger subjects, and progressively more cavalier methods have since followed. If so, one would expect abnormal psychology in children to be a growing concern. And, indeed, it is – that concern having risen with each passing decade, since the mid-1940s. Many of these children haven't acquired the logical abilities needed to make adequate sense of imposed effects, and the conceptual environment in which they live remains disingenuous.

Back in 1944 the Austrian pediatrician Hans Asperger (1906-80) coined the term autistic psychopathy; a label he applied to children who exhibited, amongst other things, a lack of empathy, clumsiness, inability to form friendships, coupled with an innate predilection for social isolation. Interest in his work increased dramatically in the closing decades of the 20th century, with **autism** (as it's now known) and **Asperger's syndrome** (eponymously named after the above) becoming just two of a whole series of psychological disorders affecting children – afflictions, which taken together, are termed pervasive developmental disorders (PDD). The very nature of child psychology predisposes youngsters to ill-effects; making them more likely than adults to develop such conditions.

Some might argue that women would be superior at combining cold clinical judgement with emotional content when dealing with children at the neuro-cognitive level. Radiofrequency technologies are capable of imposing severely on young minds, leaving them contaminated with unwarranted interference. Destructively applied the result is a pervasive disordering of their development. Therefore, putting this phenomenon on an appropriate footing is a responsibility we mustn't abnegate. In the final analysis, the numbers affected by these issues today is small as a proportion of all who will be affected; which is essentially everyone who is yet to be born. Avoiding immeasurable trauma to the same, must be our priority.

- Epiphenomenalism

Remote interference is possible because of the physics and chemistry arising at, or around, the molecular level. A combination of **intramolecular forces** (specifically, covalent bonding) and **intermolecular forces** (namely, dipole-dipole interactions, H-bonding, dispersion forces and salt bridges) both provide for large organic molecules, with strong 3-dimensional structures, coming together to produce larger bodily tissues and organs, via the ubiquitous cell. In doing so they produce person-specific absorption and emission spectra at wavelengths way beyond the visible. Conceivably, the brain is absorbing and emitting within itself, sufficient to give rise to conscious self-awareness and the presence of a soul.

Epiphenomenalism interprets conscious self-awareness, and even the soul itself, as being mere side effects of physical processes in the brain. Chemistry and physics are perceived as dictating the true nature of things, with psychology very much subordinate to the same. If true, the implications of having others tamper with those fundamental physical processes becomes apparent. You, for example, are no more than whatever those heavily manipulated physical processes happens to give rise to. Moreover, epiphenomenalists see this as an entirely one-way process, with the body affecting the mind, but the mind not noticeably affecting the body. So you would be helpless when it comes to stopping yourself from being what others, or simply your own chemistry, decide.

Advocates of epiphenomenalism are drawn inexorably into arguments about the importance of determinism and predestination in human societies. One can avoid becoming pessimistic by perceiving the whole debate as something of a dynamic. The forces of the political far Right and radical Left were sufficiently oppressive that agentic conformity and learned helplessness invariably followed. Had those political extremes been equipped with neuro-cognitive powers (remote manipulation, mental telepathy and the ability to superimpose people) they would have reduced free-will to something largely accidental. The political Centre, by definition, respects those chemical and physical processes which give rise to you – providing, as far as genetics and biology will allow, for freedom of choice.

The powers behind telepathically-induced effects have spent the last one hundred years concentrating such powers (be it informational, expert, coercive, neuro-cognitive, etc) in the hands of those who aspire to occupy the political middle-ground. That concentration of power guarantees personal freedom at the most fundamental level. Any instance in which one's physical and chemical processes aren't being respected is deviant. So much so, that if those wielding such powers were to abandon the political Centre, 'epiphenomenalism' would be our fate – our minds becoming mere adjuncts of other people's hidden agendas. Fashioning the correct constitutional framework is therefore of unparalleled importance in the present age.

That global framework remains unfinished business, with much work still to be done if telepathically-induced effects are to be safely assimilated. The politic behind these effects has outlived the extremes of Left and Right – both of which went into terminal decline not knowing how to impose on the minds of their opponents. But the male-dominated politic responsible for their decline has since become prey to its own internal imbalances and excesses. Creating the right political climate and assimilating such technology wisely would safeguard the interests of both children and society. Not doing so, however, will result in a pervasive disordering of both society and its values.

- ### Behaviourism versus the alternatives

Epiphenomenalism is just one of many closely related philosophies: such as **atomism**, in which everything can be reduced to atoms (or, in the modern age, quantum mechanics); **materialism**, which sees the human soul, like life itself, as emerging directly from the inanimate (and so is closely bound to the same); **physicalism**, in which the only meaningful statements about reality, including intellectual processes, relate directly to physical events; and **behaviourism**, in which overt actions and behaviour are seen as the only useful yardsticks in psychology. All are variations on **mechanical determinism**, and all have quantum mechanics, atoms, chemistry, physics and physical events at their heart.

Induced effects could lead to mechanical determinism, through wholesale tinkering with base-chemistry – particularly if society is undereducated and unknowing, or knowing and agentic. But equally, neuro-cognitive powers could avert that fate, as has happened over the last century, with these technologies presiding over the terminal decline of both Left and Right. In reality, the relationship between the purely inorganic, politics, mental processes and personal freedom is both complex and elastic – with that very elasticity confounding those adherents of behaviourism, physicalism, materialism and atomism. Such flexibility provides for free will, libertarianism and global democracy – and even the conscientious crafting of one's own character. With this complex relationship being particularly dynamic wherever the politics supports such dynamism.

But just because **fascism** and **communism** have passed largely into history doesn't mean we can afford to be complacent about politics. The political climate still informs the physics and chemistry from which social psychology emerges, and understanding something as profound as a person's conative dimension necessarily involves deconstructing the politics impacting on that person. The political model presented in this book starts with the presumption that nothing is forbidden, other than that which is prohibited by the people themselves – and that that which is not expressly forbidden is permissible. The overarching powers should reflect this, and desist from administering effects in a contrary manner – one which suggests that behaviour must be expressly allowed.

What this invisible hand can do, however, is selectively advantage those whose limits appear reasonable when afforded complete freedom and democracy. This slow progression to global democracy – and with it the potential to critically evaluate individuals within the same – began to accelerate between the years 1913 and 1947. With the associated technological and scientific revolution coinciding with the founding of behaviourist theory by American psychologist John Watson (1878-1958). Behaviourism sought to explain human behaviour in terms of clear-cut quantifiable responses to given stimuli, and without any reference to actual thinking. The danger with behaviourism, of course, is that insufficient weight is given to the circumstances in which the state, society or person exists. Behaviour being somewhat unpredictable, due to the range of stimuli impacting upon a given individual. Clearly, if this overarching presence is to judge fairly, fundamental attribution errors must be avoided, and that implies a more sophisticated interdisciplinary approach, one which eliminates exogenous factors such as superimposition.

Behaviourist theory did at least provide for unusual, strange and absurd forms of thinking being dismissed as inconsequential beside actual behaviour. And, in fact, many professionals would rather be judged on their most considered forms of behaviour, rather than on the complex train of free-association which lies behind the same – or, indeed, the nonsense running in parallel. But this is a courtesy to the subject, not a clinical assessment of their psychology, or their circumstance. In reality, the abnormal application of neuro-cognitive powers makes the subject's behaviour anything but a fair reflection of the person. And, in extremis, the psychiatrist is left to speculate exclusively on the politics, the self having been rendered more-or-less unintelligible.

This early 20ᵗʰ century foray into behaviourist thinking hints at the West's attempts at lifting the bonnet on the deepest processes driving society in the modern age. Engineering international primacy may well have entailed the rather harsh application of effects, sufficient to bring us to where we are today. With those seeking governance from as close to the political Centre as possible tolerating a certain amount of agentic conformity and learned helplessness, provided it served that cause. Ironically, as the Cold War intensified the Western Powers found themselves harnessing mass conformity and political helplessness in pursuit of a largely predetermined future – one based upon the core principles of democracy and personal freedom. This then, was postwar America.

• **Postwar feminist backlash**

America would argue that the robust application of remotely-induced effects was necessary in order to attain primacy for the Western Powers. Primacy having been achieved through the steeling of its many professionals to their respective roles; the agitating of key persons into socially-progressive activities; and the wholesale conditioning of its citizenry into more constitutional forms of thinking. With the dark side of this growing dependency on extrasensory intrusions being heavy drug use, endemic violence, widespread crime and learned helplessness. This highly-charged socio-political environment contrasted sharply with the more orderly attempt at strengthening the hand of the political middle-ground in a manner transcending such politics. Constitutionally sound, mainstream America was failing only in its conduct.

At the heart of these American mainstream conflicts lies a constitution with merit. One perceives the phenomenon, responsible for telepathically-induced effects, as serving the American Constitution – but over the long term, and in ways which might grate on the aspirations and beliefs of many in any given decade. This clandestine power comprises all the various persons, politics and structures, which, taken together, support the widespread application of radiofrequency effects. Inherently covert, its activities have been experienced as an assortment of unusual, strange and difficult to classify, physical and psychological anomalies. The very term **phenomenon** is therefore contemporaneous with an absence of public knowledge; and is indicative of a presence which pre-dates the proposed paradigm shift.

This male-dominated phenomenon, which still exerts its influence as strongly as ever, has been active since at least the late 1940s (and probably much earlier). It comprises a political presence with an unprecedented global reach; one which has successfully concentrated a whole range of powers in the West, won the Cold War, and provided for the free world setting the future global agenda. But its internal demographic has led to excessive ill-effects; with the fate of the individual, children especially, contrasting sharply with the overall success of the western democracies. It's not inconceivable that this male-dominated demographic has been allowed to run-on in this way in order to demonstrate the importance of women in politics – and its politics especially. A case of serving the constitution, but in ways which might appear *prima facie* unconstitutional.

This seemingly unconstitutional behaviour has resulted in a feminist backlash, with women viewing the course of events with a certain amount of trepidation. That alarm having accelerated following the end of the Cold War, when the powers behind such effects, having exhausted their original remit, appeared to be drifting. This apparent lack of direction coinciding with the rise of the cult of the living doll, an explosion in internet porn, together with a widespread exaggeration of the deeper imbalances of power within society.[22] Reluctant to resolve these issues itself, lest it lead to an idle electorate and growing incompetence in government, it's left mainstream society to correct these growing concerns.

This phenomenal neuro-cognitive power has succeeded in transcending national and subnational politics – politics which, in the West, comprises an executive, legislature and judiciary; all of whom serve to define the limits, ostensibly with the approval of the electorate. Supportive of democracy, free will and libertarianism the phenomenon's aim was to strengthen the hand of the West. And, strengthen the West's hand it most certainly has. But whilst strengthening that hand (sometimes at the expense of hard-pressed conformity and induced helplessness) it never stopped providing for all things. The unimaginable not only remained inherently possible, it actually became ever more conceivable through time.

• Neuro-linguistic programming

Psychologist Jean Piaget, the foremost pioneer of **constructivist theory**, saw personal development as the cumulative acquisition of logical abilities, taught skills and language acquisition, which together allow us to make increasing sense of the world. This applies as much to society as a whole as it does to personal development, and so a re-tuning of the broader conceptual framework will serve humanity in the long term. Only by adding to society's capacity for logical enquiry can society make adequate sense of itself and the world it inhabits. Not making sufficient sense of the world is far more dangerous, and leaves people open to exploitation. Heightened comprehension will therefore make the world a much safer place in which to live.

Moreover, beyond the mechanics of individual quantum involvement we meet with a compulsive examination of our own humanity – via the media, arts and literature. Continued deception will burden society with attributional errors, sufficient to play havoc with society's self-schema. Only the most superficial recanting of history will prove accurate under such circumstances. To avoid this we need to provide for more analytical forms of history, not simply the descriptive variety. Mass incomprehension doesn't sit comfortably with supreme interconnectedness, emergent neural networking, and multilateral telepathic involvement. Having successfully lifted the bonnet on the deepest processes driving modern civilization, and having tweaked that particular engine, so much now rest on harnessing that power responsibly.

A continued absence of public knowledge will place limits on just how complex, developed and advanced civilization can become. If we are to engage in multilateral telepathic communication; the implanting of efference copies into highly-skilled workers; not to mention, the psychokinetic manipulation of airline pilots in moments of crisis, we are going to have to lighten-up as regards issues of personal privacy, and frame their use wisely. After all, harnessing these extrasensory influences productively will take civilization to ever greater levels of sophistication, and open up a whole new world of possibilities. It's a telling reflection on humankind that we are about to make that transition smoothly and wisely. Although something unwise remains a possibility.

Society is more than a series of inanimate systems, in much the same way that life is more than a complex array of atoms; both have a dynamism which informs their constitution and structure. The study of society is called sociology, which is often subdivided into **macrosociology** (which studies society as a whole) and **microsociology** (which looks at small groups and individuals). One might expect microsociology to suffer in a world obsessed with macrosociological competition. One way to map the various trends and respond to any problems is through **sociometry**. Sociometry is able to account for personal relationships within groups and the expressed preferences of individuals, sufficient to understand the effects of long-term exposure to telepathy, and respond to the same.

One of those fields which has the potential to blur the boundary between sociology and psychology is **neuro-linguistic programming**. Neuro-linguistic programming marries together psychology, language and behaviour, in much the same way that Gestalt psychology marries together behaviour, perception and brain structure. At its most effective neuro-linguistic programming shapes perception for the better, making the person feel happier and more positive. This book is an example of neuro-linguistic programming, in-so-much as it aids comprehension, sufficient to leave the reader feeling more positive, confident and in-control. Shaping perception is of prime importance if we are to make that transition to full public knowledge decisively and effectively – free of despondency and fatalism.

Self-actualization, by which we mean making real one's talents and abilities (and using them to the full), is hampered by continuous ill-effects and the casual involvement of others. Such problems, visited on the young at school-leaving age, hints at a subduing of the body politic or a particular **cohort**, i.e. persons of the same age who are likely to share a range of common experiences. This could, of course, reflect a willingness to tolerate a measure of conformity and helplessness in pursuit of the wider macrosociological agenda. Equally, it might be seen as amplifying those innate differences between youngsters, prior to them colouring the social, political and economic landscape with their talents and abilities. Much depends on the underlying hidden agendas – and, dare I say it, the political demands of the given situation.

Generations of individuals have borne the weight of these effects, and they have done so at a time when the stakes couldn't have been greater. In the final analysis it's about consolidating the wider strategic gains, whilst preserving the rights and freedoms of the individual. Complex highly-developed civilizations are built on the judicious use of power, not power-avoidance. Critically, it's the issue of legitimacy which is all important, especially in respect of incomparable powers capable of encroaching directly on the human spirit. Future civilizations will need to call upon extreme powers in order to consolidate the global framework which is presently taking shape, and to provide for all its various systems. But if these systems end-up imposing on the human spirit illegitimately, we'll end up asking ourselves whether it was all worth it.

• Hidden agendas

Strategic concerns aside, the cold-blooded application of neuro-cognitive effects has brought with it some hard lessons and widespread discomfort. Perhaps the most significant lesson, and the most uncomfortable one if you're a nationalist, involves the need for a partial transfer of power to a **supranational body** (one which transcends national boundaries). Bearing in mind that a given state doesn't necessarily command such powers at present, it might actually end up accruing a certain amount of influence through its involvement with the same. As for direct knowledge, that is best reserved for those with a full-blooded commitment to the legitimate use of power; and it would appear that care has been taken to discredit the political extremes prior to its assimilation by those occupying the political middle ground.

Extrasensory effects exist in secret today due to the combined multinational efforts of Bohr, Einstein, Feynman, Bethe, Oppenheimer and Dyson. Therefore, the phenomenon responsible for these anomalies has been nominally supranational ever since its inception, due to the influence of Danish, German, Swiss, American and English scientific expertise – albeit expertise played-out on American soil. As for Heisenberg, it's entirely conceivable that he met and argued with the biggest names in physics, whilst still only in his early twenties, because those celebrated individuals needed him to believe the emerging orthodoxy. And that by helping to reinforce his preconceptions, he then returned to Nazi Germany to work on the fission bomb, wholly unaware of the alternatives.

Powers tend to decide what it is that they want (then work out how to get it using the powers at their disposal). And so it is with the Western Powers, who've used all the powers at their disposal to bring us to where we are today. So don't be surprised if you find evidence of **indirect rule**, in which the instruments of a given state have been co-opted through effects visited on their officials; **neo-mercantilism**, in which a commercial advantage has been gained through the manipulation of trade and industry; or **planned serendipity**, in which a carefully engineered outcome has been made to appear fortuitous or accidental. The fact that we are well placed to assimilate unimaginable powers in both a safe and progressive fashion is testament to actions taken by others over the course of a century – actions which may have involved one or more of the above (including the common denominator, **denial**).[23]

The phenomenon has been using its neuro-cognitive capability to directly influence the concentration of expert power, reward power, economic power, political power and military power ever since the early 20th century; and there's every reason to believe it should persist. In fact, the origins of civilization date back to a time when agricultural surplus grew to the point where it could be traded in a profitable manner, leading to an unprecedented degree of wealth and bartering power in one particular place.[24] Although rapacious militarism, common to both the political extremes, is also able to concentrate similar amounts of wealth, the only acceptable way is through trade and commerce. With such wealth, however, comes increased power and influence – but not necessarily legitimate forms, applied for the greater good.

Morally, the question is whether a supranational presence, transcending all national boundaries, should be at liberty to selectively advantage those, who, by their own hand, show themselves as suitably commendable – without actively denying people the freedom to be deplorable. Legitimately applied, there's no inherent reason why it shouldn't use the powers at its disposal to consolidate social progress, and provide for more of the same overseas. Effectively stateless, this overarching presence would encourage the concentration of power in the right hands, without ever assuming that the personalities of states are immutable. Moreover, it would be an open secret that those states which have gained strategic power will have to work hard at remaining legitimate in order to retain it.

Ch8:   Comprehension and familiarity

•   The future separation of power

Rousseau anticipated liberal democracy when he argued for the direct and continuous participation of all citizens in political life.[25] At its simplest democracy is just that, the ability of all eligible citizens of full-age to participate in political life, whether through voting, campaigning or standing for election. The state possesses enormous powers which the elected government is best placed to invoke, but doesn't necessarily invoke very easily. Access to the elected government is therefore access to the machinery of decision-making. This decision-making machine is headed by either a Prime Minister or President, who is constrained to act within the constitutional arrangements set-out by the state. The state, through that constitutional structure, is actively manipulating the ease with which its own powers can be invoked and the potential consequences should it choose to do so.

The present constitutional frameworks in both Britain and America demand a separation of power between the **executive**, **legislature** and **judiciary**. This means that power doesn't concentrate unduly in the hands of the executive (Prime Minister or President), but is shared with the legislature (Parliament and Congress) and the judiciary (Supreme Courts). When thinking about power it is important to understand the extent to which the constitution constrains, concentrates and fragments such power. The eligible citizen of full-age, actively engaged in political life, might end up as the elected representative of a weak government, or protesting about the actions of a strong government from a position of weakness. Much depends on the constitution.

The constitution does much to preserve the most basic rights and freedoms of all citizens, including those who take no active part in politics. This is because **decision makers** often find that power is denied, re-distributed and controlled. As for those who are politically active, the **pluralist model** of democracy sees the democratic system as providing for a near infinite variety of interest groups – from political parties, to small-scale pressure groups, charities and private individuals. In reality, participation is heavily influenced by social class (strongly related to wealth), age, gender and ethnicity. So much so that **elite pluralism** perceives power and influence as unevenly distributed. This uneven distribution of power and influence obviously affects the character of the state.

89

According to some sources a quarter of the British population possesses more than three-quarters of the country's wealth. Wealth being defined as property which can be sold and turned into cash for the benefit of the owner. And so the poor, unemployed, disabled and mentally ill all risk struggling to be heard amid potentially growing inequality. The ill-balanced application of telepathically-induced effects risks further exacerbating this trend – leading to disparities in psychological health and well-being, as well as increased impoverishment and delinquency. For individuals, societies and states to be fairly appraised the underlying machinery of democracy must be provided for, and in no way undermined.

Increasingly, Britain and America seek governance from as close to the political Centre as possible, to the extent that mainstream political parties tend to differ principally in the details. Curiously, we may be witnessing politics converging as wealth and income diverge – the political equivalent of **stagflation**. Stagflation occurs when wages and prices increase during a time of economic slowdown; the very antithesis of the Great Depression of the 1930s, which witnessed precipitate falls in wages and prices. Thus, politically, we may be witnessing rising inequality amid the convergence of the foremost political parties. Whether this cements inequality or remedies the same remains an open question. The microsociological dimension (namely, small groups and individuals) may be suffering in the name of much larger macrosociological trends and aspirations.

We've seen how the decision-making machinery of government can be strengthened or weakened by the constitution. Similarly, those competing factions seeking to influence government can be fragmented by their own internal differences. This has been called the **fragmented elite model**. The fragmented elite model is given further weight and credence by the fact that it is never wise for one faction to wholly dominate a particular political party, group or organization. The presence of sub-groups with differing opinions enables the organization to re-group around an alternative if the fortunes of the prevailing faction wane. This is often the case within political parties, whose internal allegiances are seen to shift dramatically when mauled in a general election.

This apparent inefficiency and dissension within the democratic system, together with the first past the post **electoral system**, actually serves to disadvantage the political extremes, and might therefore amount to a heavily disguised blessing. Much confusion still surrounds the ultimate distribution of power within western democracies, with sociologists very much at odds over the finer details. However, to avoid analysis paralysis I propose to adopt the fragmented elite model as *the* model of western democracy, and not cast too many aspersions on its overall reliability – objectively-speaking, it's yet to be repudiated, and some of its perceived weaknesses are potential strengths. Therefore, there's every reason to continue supporting the existing constitutional frameworks in both Britain and America, whatever the shortcomings of the politics they both catalyze and contain – those shortcomings deriving from factors other than the constitution.

The re-distribution of wealth remains a thorny issue, one which is endemic to all capitalist economies. But this issue is just as likely to be resolved within the present constitutional arrangements as any other. Moreover, the mainstream political environment places limits on what is likely to be done to those who fail to participate in politics. This affords the grass-roots of society a basic level of protection from conventional forms of power. Additionally, public knowledge of radiofrequency effects would further help to preserve the fundamental rights and freedoms of the most disadvantaged in society, affording them an opportunity to compete on more equal terms with more affluent others.

When referring to **mainstream politics** throughout this book, we (that is, myself and my heteronnubial co-author) are actually referring to the current national and subnational bodies, structures and instruments (such as the executive, legislature and judiciary), which together provide for an assortment of fragmented elites competing for power and influence up to state level. Neuro-cognitive powers aren't directly accessible by these mainstream elements. That's because mainstream elements are to be afforded a completely free hand – the only strictures being their own. Unlike the overarching supranational presence, controlling telepathically-induced effects, which would be bound by a tightly-worded protocol, prohibiting it from corrupting the integrity of mainstream democratic politics.

With these parameters squarely fixed, society can then seek to resolve its many issues and assimilate telepathic effects within the confines of the present political systems. Science acts as a portal to neuro-cognitive powers, in much the same way that mainstream politics is about access to the machinery of decision making. The phenomenon has been a shadowy presence, but revised and made public it would become a legitimate power. I propose to name this suitably revised presence **NEURON** (which is simply an abbreviation of the terms neurological and organization; but equally, it could be an acronym). Therefore, with the constitutions of both Britain and America intact, and with mainstream politics proceeding as normal, NEURON would appear out of the ether as a distinct supranational presence.

Figure 16: The future separation of power

Divisions:	Political	Military	Scientific
Supranational	NEURON	NEURON	NEURON
National	Intergovernmental Organizations and Federal Bodies	NATO	International Scientific Community
	Trias Politica	National Armed Forces	'Normal' State-funded science

The above separation of power between three major divisions (namely, the political, military and scientific) may have its corollary within NEURON itself – with scientists heavily influencing all three.

This brings us to the future separation of power (see Fig. 16). One envisages power as being classified as political (i.e. political, economic and ideological powers), military and scientific. Nationally, political power would rest with mainstream decision makers, as per a given country's constitution; military power would rest with the North Atlantic Treaty Organization (**NATO**), and supreme scientific power would be applied by NEURON (NATO having appropriated the commanding role of safeguarding NEURON, both now and in the future). In many respects this is a zero sum approach, in which access to these respective powers is fixed, rather than variable, and where NEURON's strategic gains are comparable to the mainstream's strategic losses. Mainstream decision makers (e.g. congressmen, parliamentarians, judiciary, police, etc) would have the power to make decisions and you would have the power to comply, but neither of you would have direct access to NEURON.

NEURON would oversee the politics of its member states whilst protected by NATO; which would become the world's first military force with a recognized telepathic potential. NEURON would combine having a global reach with a strictly limited remit; one which didn't intrude unduly on domestic politics. Implicit in this arrangement is the notion that mainstream politics proceeds as normal, and that matters of state are best left to the state in question. If domestic politics did deviate, NEURON would advantage those seeking to occupy the political middle ground. But if no one possessed such resolve that state would continue to deviate. The people who stand to gain most from NEURON's presence being those with the right kind of spirit and determination.[26]

This trade off between global reach and limited remit is necessary to preserve the integrity of domestic politics, whilst giving international politics and diplomacy continued meaning. Rogue states won't simply vanish from the map of the world, and might conceivably flourish if the right people don't present themselves via orthodox democratic routes. This is where electing the right people becomes critical, and why preserving the present constitutional arrangements is so important. The more commendable would be rewarded with special capabilities – but what is commendable needs to be objectively judged. Often, progressive-minded individuals, such as civil rights activists, fall foul of the law in their own country – but that's no reason not to afford them extrasensory advantages.

- **The imposition of western modernity**

The definition of **postmodernism** which has the greatest value, in this context, is that the age of grandstanding political ideology is behind us. That all those protracted debates about Left and Right are largely historical footnotes (save and except for the remedying of the same). Some view this with a certain amount of disapproval, believing that mainstream politics has lost its edge and that we're all now faced with some nebulous future, full of contradictions and ill-disciplined creativity. Others see the political Centre as having inherited the means by which to set the future global agenda – an agenda which can promote the very best of western philosophy and values, whilst borrowing heavily from the existing democratic structures. After all, managing and controlling this wider global agenda remains an important part of having and wielding significant legitimate power.

Allied to **agenda-setting** is the process of shaping and manipulating peoples desires and expectations through the media, radio and television. Thus, controlling the global agenda means taking the initiative as regards popularizing and promoting remotely induced effects. The West stands to gain the most through working with the media, radio and television, in order to enhance people's capacity for rational enquiry on this subject. This broader neuro-linguistic programming of society would actively shape society's perception of events, leaving people happier and more positive about the future. Even if the electorate experienced disquiet, it would be clear from the anticipated separation of power that mainstream politics is to be wholly distinct from NEURON. **Reference power** may afford some individuals a far reaching influence – especially amongst the oppressed minorities.

The question remains as to whether NEURON itself should be a fragmented elite, sufficient for it to regroup around a more legitimate faction should it depart from a judicious path. Deeply indented within this most secretive of powers would be those who could manage the fortunes of its own competing factions. This innermost elite is likely to have a multinational composition – many drawn from science; and a fitting demography. Some might question whether science itself should be the portal, let alone having scientists dominate its inner workings – but in principle it's apolitical, i.e. the political paradigm governing its own actions is fixed, relative to mainstream events. Real politics arises within the mainstream political environment, making altogether different demands on those charged with dealing with the same.

Judged solely in terms of the strategic advantaging of selected factions, many such scenarios have a strictly scientific dimension. Ultimately, if no one in the political mainstream shared NEURON's rather elevated opinion, they'd have no allies in the mainstream environment to advantage – mainstream politics being very much the final arbiter. If that resulted in an apocalyptic outcome and attendant population bottleneck, so be it. But at least it would be the people's own choice.

•   Supranational, not supernatural

Supranationalism implies a limited transfer of power to a multinational organization, which then acts on behalf of all involved. Given that, in this instance, the principal donor (intellectually, politically and economically) would be the NATO community, would that be so daunting? It's inconceivable that the profile of women could be raised globally without recourse to such a mechanism, albeit one operating over the course of decades.[27] Our future isn't to be the modern equivalent of a city-state, but rather actively involved in overarching constructs such as the one proposed. Constructs with the capacity to shape not just European or American politics, but global politics in the long term. The proposed neurological organization would possess a great deal of expert power, informational power and reward power, sufficient to extend our influence, not reduce it.

Assimilation is the process of incorporating new ideas and technologies without fundamentally affecting the original character of that which is doing the assimilating. Funding such assimilation would be easier with global participation. It would, after all, be difficult for America to single-handedly finance the next phase of human progress (in all its intricate detail) without such cooperation; and might even precipitate ruination, similar to that experienced by the eastern bloc, if such collaboration were wholly absent. NEURON would facilitate the application of extrasensory effects within mainstream society – in areas as diverse as education, health, politics and the law. But it wouldn't be an instrument of the state, and might conceivably empower the average citizen.

In much the same way that science benefits from **double-blind controls**, in which neither experimenter nor subject is party to the working hypothesis, society might benefit from being under the gaze of a supranational presence, working in the interests of the free world – one which can objectively determine the broader scientific realities and empower accordingly. This is where science fits into the broader separation of power (so powerfully illustrated in Fig. 16). Politics, science and the military represent three major divisions, whose different paths date back to the charismatic Richard Feynman, who's own actions reflected these three distinct areas. It would be that original relationship, writ large.

It just so happens that politics, science and the military have all been widely criticized for being too male-dominated, and for varieties of self-management which habitually favour male-typical behaviour, organizing themselves principally for the benefit of men in the process. NATO is similarly perceived as marginalizing women and being overly patriarchal – effectively reinforcing the deeper imbalances of power between men and women as a result. Given that military power derives principally from male-typical behaviour, we shouldn't be too astonished at that particular inequality. However, in the case of both politics and science that imbalance is far more difficult to justify. NEURON would help to correct that particular imbalance.

Women with a knowledge of science would be well-placed to involve themselves in the remote application of neuro-cognitive effects, provided they meet the additional criteria for knowing. This would make NATO a formidable defender of NEURON's physical security, and women formidable defenders of social and political progress. Recent history suggests that science, or at least a faction within science, has traditionally involved itself in issues of political and military strategy. As science advances, and given the growing wealth of data available, that's not necessarily a bad thing. Normal science, as in the orderly accumulation of knowledge, would continue in the mainstream by way of the scientific method. Science at the supranational level would oversee normal science, and manage revolutionary change.

- The western polyarchical tradition

Polyarchy means rule by the many, which is consistent with the direct and continuous participation of all in political life – not to mention the indirect and discontinuous alternatives. These so-called **western polyarchies** combine representative democracy with capitalist economies, and a stated commitment to **liberal individualism**. Individualists perceive the individual as being of primary importance in all matters political, with liberals also stressing personal freedom, toleration and consent. The libertarian mindset is certainly individualist, but not necessarily liberal-minded. Personal autonomy is the state of mind that libertarians value the most, and a commitment to such autonomy shouldn't be confused with toleration. Most people equate freedom with autonomy, not a liberal attitude (with that autonomous self being the person's actual persona).

NEURON (like science generally) must place value and importance on the truth, a crucial aspect of which relates to innate characteristics and behaviours. Neither NEURON nor the state should eclipse the truth with their respective powers. Those states which do eclipse the truth through mass conformity and rendering factions helpless will find themselves out-competed by the western polyarchies. Communism, Islamism and despotism, whilst being inherently repressive, also deprive those in more elevated positions of the knowledge needed to ascertain who they might induct into this proffered overarching presence. As regards **corporatism** and insider groups creating **deformed polyarchies**, in which certain elites influence governments disproportionately, such deformations would at least occur under conditions of total transparency.

That wider polyarchical whole is termed the **First World**; as opposed to the **Second World**, which is communist, and the **Third World**, which is developing.[28] Although these terms are beginning to lose their currency, they are still useful in this context, because the drama presented between these pages spans the whole of the 20th century. It is within the First World that one would expect to encounter the greatest comprehension and awareness, the maximum amount of freedom of expression, and the most prevalent extrasensory abilities. Those entering the First World can expect to find evidence of efference learning, multilateral telepathic involvement, and even **telekinesis**. But by definition they'll also encounter so much vocal resistance.

• **Opposition and resistance**

Democracy is principally about national and subnational politics. NEURON might even seem remote – outside politics – only becoming an issue as the rewards of involvement present themselves. The more Machiavellian might imagine mainstream decision makers actually provoking a reaction, in order to identify which key individuals might warrant special capabilities. However, those decision makers (be they politicians, congressman, judges, etc) are likely to have their work cut-out formulating policies and behaviours which capitalize on these external forces. Ultimately, one would expect teachers to be more conscious of the effects than their pupils; for surgeons to have a greater extrasensory awareness than the bulk of their patients; and for politicians to have survived the worst, sufficient to provoke confidence in the electorate.

NEURON wouldn't be an instrument of mainstream decision makers, and this might quell unease amongst disaffected voters. Unquestionably, it would be futile to direct anger at the politician, congressman or party chairman, as they would be subject to the same forces. In fact, the more that national and subnational politics (and the states containing the same) are subsumed within this global neuro-cognitive phenomenon the more difficult it is to single-out individual nations for direct criticism. Anti-Western sentiment remains an overriding concern – but the constitutional apparatus exported by the West provides for a nation-state being a fair reflection of the will of its own people. This provides, in theory, for all things (including an element of self-repression). The question is ultimately whether the people are being actively oppressed.

The West believes that societies shouldn't be actively oppressed, but rather they should choose their own limits and define them in a well-balanced manner. Islamists have nothing to fear from the constitutional structure exported by the West, other than the will of their own people. Furthermore, not only does the suggested global constitution provide for all things – including so much self repression – it also provides for the human mind being all things as well. Thinking is non-proscribable, in-so-far as thoughts are non-punishable within the envisaged democratic construct. So that NEURON provides for all manner of thinking and every conceivable behaviour, with only the latter being constrained at the national and subnational levels (through social norms, expressed values and the law).

Moreover, the definitive proof of thinking lies with NEURON, all other evidence is hearsay or otherwise legally inadmissible. Clearly, the mainstream cannot rest on its laurels in such a climate, as it needs to anticipate and respond to the ramifications of such freedom. This is the point at which it pays to have a phenomenon which is supremely well-balanced, wholly apposite in its effects, and with the right demography – otherwise its risks overburdening the mainstream with induced stress. The principle that the mainstream should manage itself is the right one. As is the notion that neuro-cognitive effects should do no more than determine the truth and advantage those who are well-balanced. But in reality, the wholesale destruction of both the radical Right and extreme Left has resulted in a phenomenon which is capable of over-stressing the victors.

What it most certainly isn't, is an instrument of state oppression. It is a presence which wholly transcends the politic which produced it, presenting us with both a problem and an opportunity. It can't afford to obliterate the truth regarding individuals, societies and states if it is to make meaningful use of its unparalleled powers and capabilities. And, indeed, those people, cultures and nations can't afford for it to be anything other than comprehensive, all-inclusive and global if it is to avoid them becoming hapless victims of discredited self-interest, historical bad luck, or natural catastrophe. Conceivably, it could colour history with a certain credo – providing for the spread of democracy, personal freedom, female emancipation and the scaling of new intellectual heights.

But even without it becoming an instrument of state oppression, and with the terrorist threat very much reduced through wholesale legitimacy and increased openness, one is still faced with individual distain for intrusive forces. Self-consciousness, lack of privacy and unwarranted surveillance are the most likely complaints, even amongst the most liberal-minded. Ultimately, it's a case of balancing the rewards against the more punishing side-effects. If the rewards are great enough people will overcome their reservations. Besides, punishing effects have the capacity to steel individuals to their respective roles, making them more effective and resilient. That resilience could, after all, strengthen the hand of the free world.

And, although opposition and resistance may still arise – by the same token, so might support.

•   Common complaints and popular misgivings

An individual is likely to respond initially with resentment at being known telepathically, leaving them feeling tense and resistant. Such resistance might then increase due to ill-treatment, leading to anger at externally imposed changes to habits and behaviours. These inner conflicts are simply a reflection of forced multilateral involvement, with the actions of others seeming at odds with received norms. Depending on the personality of the subject, and the magnitude of the effects, the individual might succumb to a sense of fatalism and predestination, leaving them feeling helpless or agentic. Equally, such a person may avoid regular cognitive processing, preferring a kind of self-imposed extroversion or vagueness of thought. Those inducted in this way are then faced with the prospect of re-orientating themselves, with or without the assistance of others.

This is the point at which perceptual processes, the prevailing paradigm and learned abilities become all-important. Much has been said and done to reinforce certain assumptions about the brain, and people are inclined to believe those who simply strengthen the assumptions they have. But this leaves many individuals vulnerable to destructive forms of interference, applied in ways which fall outside their personal comprehension. Before jumping on the band wagon of thinking such effects should be punitively applied, possibly in some cruel and unusual manner, remember that fundamental attribution errors are endemic to all societies and you'd be aggravating an injustice if you further harmed a victim of circumstance.

Even with the benefit of public knowledge, and with the effects legitimately applied, we're still faced with so many popular misgivings and common complaints. Perhaps the most common being **audience effects**, in which routine actions and behaviours become repressed or altered through a combination of embarrassment, self-consciousness, and a subconscious desire to appear acceptable to others. An aspect of which is **sublimation**, in which sexual impulses surface in non-sexual ways. Some might call it **social facilitation**, in which one is rendered more focused and effective by being in the company of others. Performance might well improve, albeit with the addition of a few grey hairs, but in the present climate any such benefits are likely to be swamped by excessive intrusion. This is why getting the balance right is so important.

The term excessive implies very intense forms of abuse; whether it be violently induced tremors, the impression of cerebral haemorrhage, or simply the illusion of one's stomach being cut. British humourist and writer Israel Zangwill (1864-1926) wryly observed that a pagan is what you get when you scratch a Christian. Fortunately, it has been mostly the inhabitants of Christendom who've been scratched in this manner, giving rise to predominantly bland forms of paganism. Conceivably, the doctor could know the sensations of their patients, husbands the labour pains of their wives, and mainstream decision makers an existentialist crisis or two. Such experiences might actually prove instructive, edifying or educational.

In all cases the effects must be proportional to the rewards, with punishment confined to the mainstream. The capacity of the mainstream to punish is such that it needn't be supplemented in some cruel and unusual manner. Besides, no foreign power is going to surrender its nationals to anonymous maltreatment by others (on the rather flimsy pretext of justice). What would be provided for is the psychological conditioning of key individuals, i.e. steeling them to the emotional demands of their respective roles, focusing their attention and rendering them more effective in areas of critical importance. The right person subject to pertinent effects, in just the right timeframe, might suitably excel – sufficient to anticipate the worst and respond to the same.

Ch9:   **The pathological impact**

•   Neuro-cognitive power and the rise of al-Qaeda

Soon after the kingdom of Saudi Arabia was founded, in 1932, American prospectors discovered considerable reserves of petroleum in the country. These fields were quickly exploited to feed the growing demand for oil worldwide. Consequently, petroleum exports rose steadily, and by 1970 Saudi Arabia could boast being the world's top producer and exporter of oil. This longstanding relationship between Saudi Arabia and the United States led to the Saudi kingdom being generally pro-American and anti-communist. The only thorn in the side of this otherwise productive relationship being the ongoing **Arab-Israeli conflict**, and the associated political and cultural differences. Israel, by a twist of fate, had been established in 1948, during one of modern history's most pivotal years.

Ever since the establishment of Israel, as an official Jewish homeland, there have been periodic escalations in violence between Arabs and Israelis – as happened during the **Suez Crisis** (1956), **Six Day War** (1967), **Yom Kippur War** (1973), first Israeli invasion of the Lebanon (1978), second Israeli invasion of the Lebanon (1982), first Palestinian **Intifada** (1986), and the second Palestinian Intifada (2000). All have served to strain relations between the West and Saudi Arabia. None more so than the 1973 Yom Kippur War, in which Saudi Arabia's neighbour, Egypt, launched an offensive aimed at retrieving Arab lands lost to the Israelis in 1967. This war resulted in no significant territorial exchanges, as the **United Nations** intervened, forcing concession on both sides. The real significance of this war, however, lay in its impact on the price of oil.

Saudi Arabia had withdrawn its oil from the world market in protest at Western support for Israel during the war. This action had the effect of increasing the price of oil worldwide, triggering the 1973 **energy crisis**. As oil prices rose the revenue, which then poured into Saudi Arabia, increased enormously. The real winner of the Yom Kippur War was therefore Saudi Arabia; which was able to channel its new found wealth into grandiose development projects, masterminded by foreign contractors. Saudi Arabia today is a picture of comparative modernity, albeit one straining under the weight of its highly-conservative patriarchy. This strengthening of its western ties has led to the alienation of many of its Muslim

101

youth, and divided opinion amongst the wider population – in this, the home of Islam's holiest sites; namely, **Mecca** and **Medina**.[29]

There are estimated to be well over 40,000 Americans presently working in Saudi Arabia – a significant number, but still far fewer than the numbers employed there in the boom years of the 1970s and early 1980s. Not being a democracy means that all such social, political and economic re-adjustments are more or less imposed on the Saudis, with or without their consent. These mandatory changes included the arrival of American military forces in 1990, in response to Iraq's invasion of Kuwait. Thus, many young Saudi males, who might have made careers for themselves within domestic politics, had that been an option, chose instead to 'find themselves again' fighting the Russians, in that protracted theatre of war known as Afghanistan.

The Russian's 10-year occupation of Afghanistan began in 1979, and in its own way was an attempt by the Soviets to decisively replace outdated Islamism with sexual equality and progress. Therefore, Saudi males – disaffected by rudely imposed modernity in their own country – found themselves fighting Soviet-style progress in Afghanistan, in the guise of **mujahedin fighters**. Here, in the late 1980s, **al-Qaeda** was born to organize their movements. And, following the eventual Soviet withdrawal from Afghanistan in early 1989, and the subsequent collapse of the USSR later that same year, al-Qaeda began to pose an ever greater threat to the West, and America in particular. America's permanent military presence in Saudi Arabia proving particularly provocative.

Former mujahedin fighter – and later the FBI's most wanted man – Osama bin Laden (1957-2011), claimed to have masterminded the notorious **9/11 terrorist attacks**, which took place in America in 2001. His father, Mohammed bin Laden (1908-67), had successfully amassed an immense fortune thanks to a construction business he'd shrewdly founded at the time of Saudi Arabia's creation. Such was the wealth he accumulated that the bin Laden family were able to forge close ties with the Saudi Royal Family, including King Faisal (1906-75), who was instrumental in withdrawing Saudi Oil from the world market in 1973. However, as Saudi Royals became ever closer to the US, following the arrival of American military forces in the kingdom in 1990, Osama bin Laden and al-Qaeda became mortal adversaries of both.

Al-Qaeda was, by this time, dedicated to the wholesale expulsion of Western forces from Muslim countries. In 1993 a truck bomb exploded beneath one of New York's iconic twin towers at the World Trade Centre site. This attack was undertaken with the express intention of causing the damaged tower to fall into the other. Although the destruction around the terrorist's van was serious, it failed to destroy the towers. A few years later, in 1996, the US Air Force's Khobar Towers complex (in al-Khobar, Saudi Arabia) was bombed by al-Qaeda terrorists, killing 19 US servicemen. And again, in 1998, radical Islamists, linked to al-Qaeda, attacked US Embassies in both Kenya and Tanzania, killing over 300.

In addition to al-Qaeda there are many other radical Muslim groups, often linked to specific conflicts, for example: Hamas and Hezbollah are products of the Arab-Israeli conflict, and Riyad-us Saliheen is linked to Chechen separatism. Thus, there exists a significant number of professed anti-Western Jihadist movements. These threats exist due to intractable differences within mainstream politics; as a result of the imposition of modernity in its most conventional forms; and, because of longstanding territorial disputes. Therefore, public knowledge of radiological effects won't necessarily increase the danger associated with these groups, and might even reduce the risk. Indeed, the 9/11 attacks occurred prior to the proposed paradigm shift, and with no evidence that such a shift will increase the magnitude or frequency of such events.

The tensions behind 9/11 were therefore both real and growing, with much of that anger and frustration directed at the West, and America in particular. Neuro-cognitive effects often serve to agitate – amplifying any underlying resentment and thus making conflict more likely. The male-dominated politic presiding over the application of these effects may have forced the issue in the direction of conflict – perceiving a military solution as the most profitable in terms of future stability, both regionally and globally. The phenomenon's capacity to agitate in this manner would have increased considerably following the collapse of communism in the USSR. But, whether amplified or not, the tensions surrounding the American presence in Saudi Arabia were nonetheless significant.

• **America under attack**

The two Boeing 767s – which have gone down in history as American Airlines Flight 11 and United Airlines Flight 175 – took off from Boston, Massachusetts, shortly after daybreak on September 11th 2001. Both planes were almost immediately hijacked by Islamic terrorists working for al-Qaeda. The terrorists then assumed full control of both planes, piloting the same for the remainder of their fateful journeys. Flying straight and level, Flight 11 ploughed directly into the North Tower of the **World Trade Centre** at 8.46 am. Then, just a quarter of an hour later, at 9.03 am, Flight 175 smashed into the side of the South Tower. A combination of copious quantities of jet fuel, massive impact forces, lightweight materials, inadequate fireproofing, together with an over-reliance on trusses, fatally weakened both buildings. Collapse became inevitable as the very fabric of the towers weakened due to the presence of immense heat. Finally, at 9.58 am, the South Tower collapsed; and 30 minutes later so did the North Tower.

As the twin towers had stood burning at 9.37 am, a third aircraft – American Airlines Flight 77, en-route from Washington to Los Angeles – buried itself into the side of the Pentagon Building, in Washington DC. The resultant explosion tore a monstrous hole, in this, the headquarters of the US Armed Forces. Throughout this initial phase calls for help were still being received from those stranded in the twin towers above the two points of impact. At more or less the same moment as the calls for help from the South Tower abruptly halted, due to its unexpected collapse, a fourth aircraft – United Airlines Flight 93 – nose-dived into the ground 80 miles south-east of Pittsburgh, killing all on board. United Airlines Flight 93's passengers having heroically wrested control of the plane back from the terrorists.

We can only speculate as to precisely how destructive the terrorists expected their actions to be, given that they were flying fuel-laden commercial aircraft approaching their maximum speeds, and with only a limited knowledge of the physics. According to Leslie Robertson (1928-), the lead structural engineer on the Twin Towers project, "the energy contained in an airplane, or any other moving object, is proportional to the square of the velocity; double the velocity and you have four times the energy – and so the energy imposed on the building by the 767 was very substantially more than the energy that we had assumed in the original design".[30] In all probability, no one, not even the principal engineers on the twin towers project, could have accurately predicted how events would play-out, physically and structurally on that fateful day.

Osama bin Laden probably anticipated American morale twisting, warping and collapsing far more than the twin towers themselves – not unlike previous events in Vietnam or Somalia. In the words of Orla Guerin, a BBC Middle East Correspondent at the time: "On September 11th, the United States, so proud and so free, became a part of the Middle East – a scene of carnage and sudden death, a place of fear, insecurity and grief".[31] Her words echoing the Israeli newspaper which had proclaimed several days earlier "It is our tragedy, writ large". The wider perception was one of Americans being very much world citizens, equally vulnerable to the wholesale ramifications of mainstream politics, both at home and abroad. What I propose to do here is conjecture as to the raison d'être prevailing at the time of the incident, assuming that an extrasensory dimension was indeed present.

Conceivably, such an incident could have occurred with a supranational body observing. One can see evidence of the same in the actions of the United Nations in places such as Bosnia, Rwanda and Cambodia. All of which have witnessed people being abandoned to their fate by more powerful others, and in spite of those others knowing that deaths would occur. Americans are clearly no more immune to that aspect of history, than say, Bosnians, Tutsis or Cambodians. So, perhaps there was a supranational presence presiding on that day, one which chose to abandon several thousand Americans to an early death – begging the question as to whether any person should be abandoned in that manner, and what manner of mysterious workings might give rise to such abandonment.

Having established that people can and do get abandoned to their fate by overarching powers, we need to consider what might have motivated those with a neuro-cognitive capability to act in that way. In other words, what cold-blooded logic might the phenomenon, or even NEURON, have adopted, in the days, weeks and months leading up to incident. Clearly, the tensions leading-up to 9/11 were genuine, rather than fictitious, with much of that rage directed at America, modernity and the West. The fact that the mastermind behind 9/11 was Saudi, as were 15 of the 19 hijackers, suggests that America's unwelcome presence in the Middle East, and Saudi Arabia in particular, was the principal source of anger driving those events.

NEURON wouldn't exist to perform the jobs of mainstream decision makers, be they presidents, congressmen, politicians, or simply airport security staff. The electorate must still exercise its collective will wisely and not rely on some invisible hand to compensate for their lack of interest in domestic politics. Besides, NEURON's strictly limited remit makes mainstream politics and foreign policy ostensibly matters for the state – together, of course, with the ramifications. A neuro-cognitive power acting in the long-term interests of the free world could, in theory, leave in its wake so many unsuspecting victims of its broader strategizing.

105

Thus, we find that the West has at its disposal subnational, national and supranational powers, with the last of these being called upon in the case of more intractable global problems.

For example, if US foreign policy resulted in real and growing tensions – as with American military involvement in Saudi Arabia – America would be called-upon to foil every single terrorist plot. NEURON would take the view that responsibility for foiling every plot rests squarely with mainstream decision makers in the US. This is where electing the right people every time becomes critical – especially if the threat is ongoing and the number of plots is expected to increase through time. This danger, which results from having a permanent western presence in the home of Mecca and Medina, would have created an enormous burden on the US authorities – with NEURON taking a somewhat detached, far-reaching view of events.

By not intervening, NEURON would have allowed the failings of America's own policy makers to determine the outcome. Such intervention would, in any event, make people less accountable for their own actions, sow the seeds of growing incompetence in government, and diminish the separation of powers previously mentioned. Just as the presence of a judiciary doesn't prevent crime, the presence of NEURON wouldn't, in itself, prevent tragedy ever occurring. In the years leading up to 9/11, NEURON would have been conscious of the psychology of those abroad, sufficient to know the futility of a defensive strategy. The only practical solution, for America and the free world, would have been war on the ground in foreign climes. Ironically, it was the failings of mainstream decision makers which guaranteed this.

We might conclude by saying that wherever and whenever the free world is best served by national or subnational politics, so be it. But every now and again supranational politics takes precedence, possibly at the expense of individual nation states and even individual citizens. On those specific occasions, the only practicable solutions (the ones which are most likely to endure) transcend the whole of domestic politics. That's the reason why some would say that the perceived neuro-cognitive power acted prudently and within its given mandate. That global mandate, however, must be entirely legitimate if those sacrifices are to be adequately respected – and, indeed, justified.

- ## Interconnectedness and disorder

Some very male-typical behaviour has brought us to this point, both in the western hemisphere and in the Muslim world. Evidence arrived at over the preceding one hundred years fairly repudiates the premise of indistinguishable male and female brains. Political pathologies are invariably patriarchal, i.e. dominated by men. The **Greater Middle East** comprises, for example, so many all-encompassing patriarchal systems – ones which spawn more than their fair share of pathological behaviour. Women are more well-balanced than men, on average, and therefore less likely than men to react badly to excessive interference. Female emancipation is therefore an essential precondition for knowing radiofrequency technologies. This is the ineluctable conclusion arrived at since the turn of the 20th century, and which has been so dramatically underscored by the 9/11 tragedy.

Behind this fundamental precondition lies a conscious determination to see neuro-cognitive powers properly applied. Over the last few decades routine excesses have led to a growing catalogue of conditions, syndromes and afflictions. Psychiatry, of course, has sought to apply labels to those exhibiting discernable symptoms, taking on trust the chemistry, physics and biology presented to it. However, the arbitrary manipulation of chemical processes in the brain, at the molecular level, can bring about a near infinite range of disorders. This disordering of individuals, some no more than children, stems from the ill-balanced application of such effects. Suitably revised, psychiatry needn't undertake a far-reaching revision of its practices. Much depends on our ability to eliminate exogenous factors as the principle cause of psychological stress.

Provided that innate characteristics aren't being overwhelmed by destructive forms of interference, a judicious assessment of personality and behaviour can be arrived at. However, the ill-balanced application of such effects will cause the good to rebel and the bad to appear agentic. Worse still, the good might experience a disordering of their character and then find themselves judged by those who aren't better – just better treated. The opposing forces model (see Fig. 17) shows how an individual, subject to ill-balanced interference in an otherwise well-balanced society, might become troubled by a range of disorders. Even in the absence of such symptoms, the subject's state of mind may be coloured by excessive meddling.

Figure 17: The opposing forces model

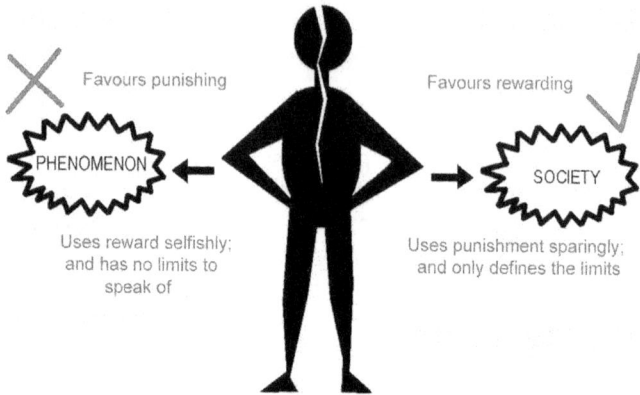

Favours punishing
✗

PHENOMENON ⟵

Uses reward selfishly;
and has no limits to
speak of

Favours rewarding
✓

⟶ SOCIETY

Uses punishment sparingly;
and only defines the limits

A crucial aspect of this debate pertains to the psychology of those applying the effects. During the war the Allies defined a partisan as an opponent of Nazism, and therefore someone who would vilify the likes of Reinhard Heydrich (one of the chief architects of the Holocaust). Now imagine, for the sake of argument, a person with the disposition of Reinhard Heydrich presiding over just one person remotely, with that person then vilifying the powers behind induced effects. That vilification may be entirely justified if the subject has been cruelly treated. Clearly, for neuro-cognitive powers to be associated with the preservation of the free world, rather than the corruption of the same, certain preconditions are called for.

What I propose to do, is argue that punishment should be the preserve of the mainstream, whose capacity to castigate, penalize and correct needn't be supplemented. In fact, there is no more telling test of logic, wisdom and integrity than the law – making criminal justice a psychometric appraisal, watched over by more powerful others. This may seem like one is washing one's hands of the whole affair – with potentially catastrophic implications for that special individual with enhanced faculties – but ultimately this strategy serves to reveal the truth about individuals, societies and people. Besides, the benefits of an extrasensory capability are such that failing to possess that advantage is punishment enough.

Those enhanced faculties would be derived from an overarching power acting mysteriously in the wider interest – one concerned with implanting efference copies in highly-skilled workers, steeling police officers to their duties, evaluating the performance of the judiciary against privileged knowledge, facilitating telepathy between NATO personnel, testing mainstream politicians with powerfully-induced effects, providing for advanced civilization by tweaking the deepest quantum processes underlying the same, and all the while adding to global peace, stability and security in the long term. Moreover, psychological illness would be less of a lottery, being more akin to problems arising in those who are failing to capitalize on the advantages of knowing, either through self-consciousness, embarrassment or fear.

• Injury, disease, chemistry and genetics

Psychiatrists and brain surgeons focus exclusively on problems arising due to brain injury, degenerative disease, biochemical imbalance and genetic predisposition. Very punishing examples of remote manipulation can simulate all four, irrespective of the person's predisposition. Banning the punitive application of such effects does at least assist clinical diagnosis and prevents a return to the bad old days of potentially unwarranted prefrontal lobotomies. Having avoided mistakenly attributing neuropathology to injury, disease, chemistry and genetics, by effectively eliminating exogenous factors, society would then be better placed to assist those with genuine psychological problems. However, even with **punishment** confined to the mainstream, problems may still arise due to neuro-cognitive influences.

Simply superimposing a conflicting normality may feel pathological. For example, dreaming might be experienced as hallucination, alertness as insomnia, tiredness as chronic fatigue syndrome, self-absorption as social autism, partial erection as implied sexual mania, and talkativeness as the first signs of madness, etc. In defence of the same it's possible to argue that induced lethargy might serve as a kind of resistance, which has the effect of increasing one's stamina. And that if one induces talkativeness, an individual might then exercise their less verbal side. Cynically applied, however, these otherwise normal aspects of daily life constitute a form of duress, amounting to so much subjugation, suppression and control of individuals.

Another example of normality producing abnormality is in respect of physiologies arising in extremis. The human body can react to extreme situations in a variety of ways. Take, for example, extreme cold. Simulating a normal reaction to extreme cold can result in induced trauma to certain body parts, principally the extremities, through a reduction in blood flow (e.g. vasoconstriction, hypovolaemia, etc). One could argue that all induced effects are nominally natural, in-so-much as they action neurophysiologies which the body could, in theory, produce itself. But these largely sympathetic and parasympathetic reactions are generally a reflection of the prevailing reality, not casual excess. After all, the whole point of having a unique absorption profile is to avoid such processes being actioned in an *ad hoc* way.

Take enzyme action, when two reactants become loosely-bound to the surface of an enzyme, distorting their covalent bonds (sufficient to accelerate the chemical processes which sustain life), it was never intended that the principal trigger would be something as alien, facile or remote as politics. It would be a mistake to supplement injury, disease, chemistry and genetics with exogenous trauma, arrived at through the willful or destructive use of remote effects. As we've seen, you are little more than what those heavily manipulated processes happens to give rise to. Just how heavily manipulated those processes become, remains to be seen. Taken to an extreme we might become shackled to a deterministic future, trapping us in an agentic role we never sought or a struggle we can never disengage from.

- ## The process of enrichment

Individual minds, like the societies which sustain them, absorb and emit within themselves, sufficient to give rise to conscious self-awareness. Such self-absorption lifts humankind above all other animals; but it also leaves us vulnerable to the interpretations we then place on what our senses are telling us. Some psychologists subscribe to the idea of **differentiation**, in which personal development enables a person to recognize the key features of a given stimulus with increasing alacrity. More probable is the theory of **enrichment**, in which individuals supplement the information provided by such stimuli with their own ideas, knowledge and insights. Thus, the theory of enrichment is more or less consistent with the constructivist theory of knowledge, whereby the gradual acquisition of logical abilities, through learning and experience, aids perception – or at least supplements what we experience in an increasingly sophisticated fashion.

One can, of course, acquire erroneous knowledge and illogical thought processes, sufficient to undermine perception – leading to mental illness, **sociopathy** and terrorist activity. Enrichment theory neatly provides for one bringing faulty ideas, inadequate knowledge and defective insights to bear, when judging the information provided by our own senses. How we develop as individuals and how we enrich the information conveyed to us from various sources all affects our interpretation of the world around us, together with our place within it. Suitably enriched, we are protected from fear, anxiety and depression. Others also benefit from this neuro-linguistic programming of our person, through them not falling victim to our misplaced anger, resentment or paranoia. A well-balanced person will always favour rewarding; using punishment sparingly, and only to define the limits. Destructive feelings and emotions arise from imbalances in our thought processes. Self-destructive feelings, for example, occur in those who fail to reward themselves with sufficient praise, compounding that failure with excessive self-criticism. Sociopathy and terrorism, on the other hand, arise when society bears the brunt of punishing behaviour, wholly out of proportion to the cause. Destructive interference, or, if you prefer, injurious psychology, amounts to an individual favouring punishing, having no limits to speak of, and using reward selfishly or not at all. Unfortunately, we all deviate from the optimal, psychologically-speaking, hence the need for power to be separated, fragmented and denied.

In the case of NEURON, the power to punish would be withheld – that prerogative amounting to a psychometric test of mainstream logic, wisdom and integrity. Not only would the phenomenon in its revised form be prevented from punishing, it would be prevented from superimposing punitive-minded behaviour. This would mean that terrorism, sociopathy and suicide couldn't be remotely induced. And, it would lessen the risk of a person, negatively enriched both psychologically and emotionally, from being exploited. By pro-actively managing the process of enrichment – injury, disease, chemistry and genetics, notwithstanding – the mainstream democratic environment can more effectively manage the wider acquisition of knowledge, ideas and insights, or at least supplement the same with sounder alternatives.

- Rethink

Chemical messengers acting singularly, or in combination, across trillions of synapses, have the effect of controlling, regulating and inhibiting the movement of signals. Nerve impulses from the eye reach the brain via the optic nerve, leading to the emission of radio waves – sufficient to excite specific groups of adjacent neurons. These groups of neurons provide a convenient basis for **memory** – the resultant image being fixed chemically.

Once fixed, that group can stimulate the source neurons using emitted energy. One's self-schema would amount to a plurality of chemical imprints, arrived at through the absorption and emission of thoughts – thoughts, which if re-absorbed enough times, will hard-wire themselves into our actions and behaviour. This is pure speculation, of course, as the underlying mechanisms responsible for memory and consciousness remain obscure. In fact, contemporary neuroscience appears somewhat Copernican in its comprehension, possibly due to the politics.

What we do know is that the home computer is heavily reliant on a central processing unit (or CPU), which, like the brain, receives 'afferent' inputs from the keyboard, enriches that information using its memory, prior to outputting an answer, in an efferent manner, to the monitor. What the computer can't yet do is engage in efference learning, manage its own memory, and develop a self-schema. Images, thoughts and ideas fixed in our minds, by way of memory, shape our interpretation of incoming information, sufficient to determine our actions and behaviour. Actions and behaviour which are anything but biologically predetermined, as so much depends on how we enrich the information conveyed to us. Herein lies the source of all that elasticity which so confounds the adherents of behaviourism, physicalism, materialism and atomism.

The West, seeking to capitalize on that underlying mutability, has engaged in a century long struggle to achieve: firstly, international primacy; secondly, a full command of the telepathic potential naturally extant within the human brain; and thirdly, the prudent application of that power. Only the last remains elusive as we strive to openly assimilate the extrasensory component. Our capacity to augment external information constitutes a two-edged sword, as the ill-balanced may fabricate self-destructive thoughts or punish society in a wholly disproportionate manner. But equally, society risks falling under the spell of deterministic forces, leading to agentic conformity, learned helplessness and the wholesale destruction of its innate characteristics if kept completely in the dark. Disorders accruing from openness are to be preferred.

Mental illness (e.g. **anxiety**, **depression**, **phobias**, **neurosis**, **anorexia nervosa**, **bipolar disorder**, **ADHD**, Autism, Asperger's syndrome, **psychosis** or **schizophrenia**) is indicative of a pernicious imbalance. Although that imbalance is likely to result from injury, disease, chemistry or genetics, induced effects may actually serve to aggravate the same, especially amongst those who are young and unknowing. And, there remains the strong possibility of patterns of behaviour being established through third-party interference. The well-balanced application of such effects would at least isolate any natural propensity to neuropathology. Ultimately, if those applying the effects are permitted to imitate nature it will become difficult to deduce which symptoms are natural. More broadly, it will become difficult to determine who the real terrorists are.

PART IV

Balanced-mindedness

Chapters 10-12

Ch10: The well-balanced mind

• Impart answer

An axiom is a rule, statement or maxim that people accept as being self-evidently true. One such axiom comes not from Heidegger, Plato or Descartes, but from childcare – namely, favour rewarding; use punishment sparingly, and only to define the limits.[32] Conforming to this rule makes one **well-balanced**. Fortunately, the molecular structure of individuals and the technological structure of democratic societies enables the absorption and emission of signals both within and between people, with memory and learning rendering our responses somewhat dynamic. Therefore democracy neatly provides for a given population crafting a well-balanced social environment – one which can achieve real efficiencies by defining reasonable limits with the sparing use of punishment. And, which can periodically revise those limits, so as to be as reasonable as possible.

Conversely, ill-balanced and unruly societies favour punishing; using reward selfishly, and without obvious limits. To subscribe consciously to this unruly alternative is to be self-destructive, malign or evil. Mentally unbalanced individuals have no clear limits; whether it be in terms of the amount of punishment inflicted on themselves (anorexia, self-harm and suicide), or in terms of the punishment meted-out to others (grievous bodily harm, torture and murder). Well-balanced societies have a duty to tackle suicide and self-harm, just as they have a duty to tackle torture and murder. The dissolution of democracy renders a country less mutable and more deterministic – compounding that effect with propaganda. Moreover, power usually concentrates in the hands of a male-dominated oligarchy.

Many of the phenomenon's present imbalances stem from male-domination of its internal structures, which are undoubtedly patriarchal. In an open competition, based upon the above **rule**, or precept, one would perceive women as having a competitive edge or advantage. Armed with such an advantage women could cure the politic (responsible for such anomalies) of various imbalances and excesses. Moreover, they'd be ideally placed to advance the cause of women worldwide, through the pertinent application of neuro-cognitive effects. This would see women more fairly rewarded than they are at present – politically, socially and economically – and it would serve to protect the interests of children.

Well-balanced societies seek to make the world less punishing, or at least no more punishing than it need be; whether through forgiveness, due process, self-censorship, medical advance, or even regime change. In the case of regime change we might see punishing male-typical military activity on the ground competing with rewarding female-typical neuro-cognitive activity across space. Such **nation-building** amounts to the establishment of representative democracy in place of despotism – drawing such countries more closely into the realms of the free world in the process. This should, in theory, increase the opportunities for individuals living in such countries, making their lives more rewarding as a result. Moreover, they'd no longer be burdened with an inflexible democratic deficit.

Balanced-minded individuals are inclined to reward those with merit, both at home and abroad – whether it be with praise, status, power, knowledge, or simply money. As regards the redistribution of wealth within society, one could argue against wealth concentrating in the hands of those who lack merit. For reward power to mean anything, one would expect extrasensory powers, like all other rewards, to concentrate in just the right hands. It just so happens that the West has set the most reasonable limits, censors the most destructive forms of rhetoric, promotes sexual equality, and enjoys a suitable separation of its powers, and is therefore more qualified than most to possess a neuro-cognitive capability (and, quite frankly, use it).

- ## Internalizing and externalizing

Freud's overarching view of the psyche, incorporating the id, ego and superego, although a little tired-looking after all these years, still serves as a useful point of reference. If the ego (self-schema, consciousness and planning) and superego (conscience, scruples and morality) result from the brain absorbing and emitting within itself, sufficient to chemically imprint the person's persona onto the very fabric of the brain, then the id (unconscious source of primitive drives and desires) must derive from the chemicals and fabric of the brain itself. That chemical-soaked 'rag' being sensitive to radio waves arising from within its many folds. Thus, radio waves mediate between consciousness, memory and self, in a manner which is highly dynamic. Unfortunately, the fabric of the brain is subject to genetic or biological determinism, and therefore predetermined by our genes, whilst the ego and superego remain relatively pliable.

Freud excelled at painting a dynamical view of human psychology, one in which immutable base-instincts vie with the more fluid conscious mind, in a never-ending battle watched over by one's conscience. Because of the key role played by radiofrequency waves, this particular mêlée can be accessed remotely by those seeking to influence personality and behaviour. This is where sexual differences begin to materialize, resulting in some further fine-tuning to the theory of male and female brains. Women being more likely to **internalize** their issues, resulting in disturbances of mood (e.g. worry, anxiety, guilt, depression, etc), and men being more likely to **externalize** theirs, resulting in affected behaviour (e.g. aggression, violence, anti-social activity, etc). All of which has its corollary in the crime statistics.

This duality of spirit is most conspicuous in adolescents, who are precisely the group who may be unaware of external influences. Consequently, youngsters are particularly vulnerable to exaggerated reactions born of remote interference – such as para-suicidal gestures, rebelliousness and anger. Some of these gestures may be 'involuntary, voluntary ones', i.e. gestures which the person themselves perceives as being entirely voluntary, but which are, in fact, driven by third-parties. Such youngsters risk incurring mental and emotional damage due to effects visited on them beyond their five senses – leaving them terribly unresolved, conflicted, and potentially suicidal. A personality, unbalanced in this way, may internalize or externalize; much depends on the person themselves.

Often the level of violence needed to diminish a person's prospects is considerable; hinting at the individual's comparative soundness. Strong people might rationalize, sufficient to progress their lives; but even then they remain vulnerable to excessive intrusion. Such incursions serving to undermine the diligent re-crafting of their character. The truly balanced mind, which owes so much to conscientious re-crafting and periodic self-analysis, shouldn't be cynically undermined by anonymous third-parties. It would be an unruly presence which sought to oil the wheels of self-destructive thinking and wider conflict. Far better that these arise due to mainstream causes; ones which can be deliberated upon by conventional analysts – rather than as a consequence of some invisible hand acting in a disorderly fashion.

It would serve society if those wielding neuro-cognitive effects chose to internalize more than they do at present, sufficient to mirror the very best of women's instincts; and not encourage the gross externalization of personal grievances in a way which is ultimately counter-productive. In the final analysis, those who are grossly-externalizing and punitive-minded are best confined to mainstream society; which American Poet Ralph Waldo Emerson (1803-82) once piercingly described as "a hospital for incurables". That incurable dimension being the unyielding part of our inner selves – compensated for, in the first instance, by our egos and superegos. In much the same way that democracy provides for the citizen shaping the personality of the state, conscious self-awareness provides for the conscientious crafting of one's own character. However, the power to do both is often withheld, fragmented and controlled – sometimes by one's own biology.

• **Normative analysis**

Objective analysis looks at how things are; conversely, **normative analysis** questions how things ought to be. Any revision of the phenomenon's practices is going to require an element of normative analysis. The word normative isn't strictly-speaking synonymous with normal, in-so-far as what one ought to do isn't always regular or routine. Normatively-speaking democracy should emulate the arrangement known as *trias politica*, or three political bodies – which comprises: 1) an executive, who formulates policy; 2) a legislature, which enacts laws; and 3) a judiciary, to decide on legal cases. One can even normatively analyze the finer workings of this arrangement, but objectively-speaking the demands of the real world often serve to create anomalies and inconsistencies at the day-to-day level.

Beyond individual democratic states, what ought to happen is NATO and NEURON binding the free world together both militarily and neurologically, with each serving the others interests. This would render the free world a coherent whole, possessing a pronounced extrasensory capability. Those nations which surrender some of their sovereignty to this global presence might find that they gain power and influence through their cooperation. It's important to stress that if you want to change the world become an elder statesman like Sir Winston Churchill – if, however, you want to help such a figure to change the world, be inducted into NEURON. NEURON being dedicated to the conservation, preservation and advance of both freedom and democracy.

In the case of regime change that extrasensory capability would be used to advantage those pursuing the pluralist cause. Even without exercising one's voice, the imposition of a modern constitution does much to protect minorities and political rivals from the worst excesses of the executive. In Robert Fisk's timely book on the Greater Middle East, entitled The Great War For Civilization (Fourth Estate, 2005), it's clear that without a progressive constitution, adequate separation of power, pluralist thinking, sexual equality and representative government, dissenters are likely to end up being "burned with acid", "tortured with electricity", or, in the case of 5,000 inhabitants of the Kurdish town of **Halabja** (in 1988), simply "gassed to death".

Seen in this light the imposition of an enlightened constitutional framework, one which prevents power concentrating demonically in the wrong hands, might not seem so bad. Ultimately, it's about the expressed will of the people; what we might term the state's true conative dimension. Conceivably, subnational politics could still infect such a state, rendering it rogue. In which case mainstream forces would be called-upon to address that malign presence. A malign presence which could, in theory, turn out to be a monstrous tyranny. But even then, provided the global framework isn't compromised by lightweight rhetoric, copious quantities of pusillanimity, or massive impact forces, our global future should be assured.

Only the destructive power of nature exceeds these political and military conflagrations; whether it be catastrophic climate change, burgeoning ice-sheets, asteroid impact, supervolcanic eruption, or mass extinction. A robust global framework, supporting a neuro-cognitive dimension, might help to absorb the very worst of such ruinous events. Episodes which are bad enough when all they appear to affect is the earth's physical climate. Today, however, these events are likely to agitate a whole array of political, military and economic climates as well. These climatological phenomena having diversified through time, broadening what might loosely be described as man's environment. The question remains, however, as to what mankind ought to be doing – both in terms of the types of climates he creates for himself, and his chosen responses to the same.

•   Anthropobionomics

Bionomics is the study of a single species in relation to its environment; and so **anthropobionomics** amounts to an examination of mankind's relationship to the whole biosphere. The world's **biosphere**, which contains all known life, comprises a number of climatic zones – namely, the **equatorial, sub-tropical,**

120

temperate and **polar** zones (with an additional region in the northern hemisphere, termed the **boreal** zone). Modern humans left equatorial Africa around 80,000 years ago, eventually venturing north into the temperate, boreal and polar regions, where they sought to maximize the rewards associated with these increasingly higher latitudes, whilst reducing, as far as possible, the punishing effects of longer colder winters and increasingly shorter cooler summers. Clothing and technology evolved far more than the people themselves, as they adapted to these new harsher environments.

All the world's ethnic groups and cultures, including those presently living in the stone age, comprise so many physically modern humans. All have the ability to adapt, innovate and invent, but none is inclined to do so for its own sake. That having learnt to survive in a given locale, progress then freezes. What appears to drive ingenuity and inventiveness isn't instinct, it's necessity. It seems doubtful that the brightest of people would spontaneously invent in the absence of any kind of social conditioning, active tutoring or obvious stress. Political, military and economic demands seem to create an impetus all of their very own, driving that aspect of ourselves which would otherwise remain dormant. These stresses becoming particularly acute around the turn of the 20th century.

By remaining within roughly the same latitude as their African homeland, the ancestors of today's Papua New Guinea tribesmen avoided having to be overly innovative (as points along a given line of latitude share the same length of day, and therefore similar climate and vegetation).[33] As modern humans reluctantly confronted the vagaries of the higher latitudes, advances in clothing and technology became unavoidable. Accordingly, modern humans would have populated the world in the manner of electrons filling a Wolfgang Pauli atom – but instead of electrons filling atomic shells in a prescribed manner, it would be the migrants themselves, filling those areas which required the least innovation and adaptation first (followed by those regions requiring some innovation and inventiveness, and so on). With this innovation came advanced civilizations.

Prior to the arrival of these highly-organized societies, some 5,000 years ago (following the domestication of both plants and animals) the demands where primarily meteorological. With the advent of civilization, however, new forces emerged, creating for the first time a variety of competing political, military and economic stresses. These disparate man-made climates were

121

more capricious, shifting and erratic than either the geography or the weather, creating demands all of their very own. Progress was unable to freeze, given the competition between nations, and these emergent states became the seedbeds of innovation. Today, every nation-state seeks to maximize the rewards associated with these rather abstract climates – whilst reducing, as far as possible, the more punishing effects associated with the same.

And so it was at the turn of the 20th century, when Germany was determined to compete with both Britain and France overseas, fashioning itself into the guise of a major imperial power in the process. The eventual containment of Germany, half a century later, required, as we've seen, the combined weight of Britain, France, Russia, America, and all their allies. Contemporaneous with this global power struggle was the invention of telepathically-induced effects, whose presence created an entirely new zone. Those who enter this new, some would say hostile, neuro-cognitive zone, will no doubt seek to maximize the rewards associated with the same; whilst reducing, as far as possible, its more punishing aspects. After all, entering new and challenging regions, reluctantly or not, remains a quintessential part of what makes us modern humans.

- **Without victory, there is no survival**

Fate can't be relied upon to be a well-balanced overarching presence. Chance, providence and even abject freedom can all too easily lead to the likes of Nazi dictator Adolf Hitler (1889-1945), Soviet despot Joseph Stalin (1879-1953), or Cambodian tyrant Pol Pot (1925-98). And so the ability to assist fate in addressing or avoiding such hazards is clearly important. Those applying the effects might argue that such tyranny was natural or innate; but more can still be done to assist those challenging the same. After all, those who confront such despotism in the mainstream, without the safeguard of anonymity, are far more deserving of extrasensory powers than the personnel providing for the same. However, assisting individuals, particularly in far-flung theatres of war, may have been difficult prior to the postwar era.

One believes that the human brain's latent telepathic potential was unlocked sometime between 1913 and 1947. Too late to have affected all but the most significant figures in those years. Remember, the economic, political and military climates of the 1930s and 40s were some of the most inclement ever faced by humans. Consequently, innovation, ingenuity and inventiveness were stretched to their limits. As, indeed, was our capacity for violence. In the event, it was an economic storm which set in motion a train of events which would unleash mankind's deepest, darkest and most unruly impulses – the subsequent conflagration catalyzing weapons of such supreme ferocity they can only be trusted to those who fragment power, distribute it wisely, and who leave the executive questioning their status. As the postwar political environment, so central to subsequent chapters, emanates directly from that particular tempest, we must briefly reprise that history in order to critically evaluate the psychology of some of its foremost participants.

In 1929 the **Wall Street Crash** caused the flow of capital out of America to cease; leading to the collapse of the German banking system. Severe reparations payments, coupled with disarray within the German financial markets, and a mounting communist threat to the east, all helped to make Adolf Hitler's Nazi Party the single biggest party in the German Reichstag. Having nominally achieved the position of chancellor, Hitler then called fresh elections, polling 43.9% of the vote. Shortly afterwards, an Enabling Act (1933) gifted Hitler absolute power. Hitler's executive powers were swiftly used to fatally compromise his political opponents, and later the Jewish minorities. This wholesale terror leading inexorably to agentic conformity and learned helplessness amongst the remaining masses – democracy having, in effect, liquidated itself.

Hitler then embarked on a gross externalization of his deepest grievances: Austria, Czechoslovakia, Poland, Denmark, Norway, Belgium, the Netherlands and France all falling in the face of Nazi aggression. His Nazi regime had no limits to speak of, both in terms of the cruelty meted-out, and in terms of the benefits it selfishly appropriated – making it, quite simply, pure evil. That evil took a new and sinister turn in the summer of 1941, as Russia entered the war on Britain's side, following Germany's invasion of the Soviet Union. The sheer scale of anti-Jewish killings escalated alarmingly in the east, as the Nazi German invaders began to target men, women and children.[34] Contemporaneous with this move to mass extermination was the requirement, in Germany, for Jews to wear the **yellow star** – in anticipation of their forced deportation.

Thus, the 1942 **Wannsee Conference**, which was chaired by Reinhard Heydrich, one of the Holocaust's chief architects (mentioned in chapter 9), appears to have done little more than hammer out the finer details of how that anticipated slaughter of European Jews was to be achieved. And the manner by which it was to be achieved, it transpired, was through identification, deportation and gassing. Later that same year extermination camps were established at Treblinka, Sobibor, Majdanek and Belzec. The long since established concentration camp at Auschwitz (the largest of all the camps) doubled as both a labour camp and as a place of mass-execution. All these death camps were situated inside Nazi-occupied Poland.

Sir Winston Churchill in a Common's speech had said: "You ask, what is our policy? I say, it is to wage war by sea, land and air; with all our might and with all the strength that God can give us. To wage war against a monstrous tyranny, never surpassed in the dark and lamentable catalogue of human crime. That is our policy. You ask, what is our aim? I can answer in one word, it is victory. Victory at all costs. Victory, in spite of all terror. Victory, however long and hard the road may be, for without victory, there is no survival". In this clash of titans, one the personification of pure evil, consciously subscribing to an unruly use of power; the other equally determined to mete out punishment, sufficient to define civilized limits in the world; it would be the latter who would warrant an unnatural advantage.

One person who's spirit shone brighter than most in those dark and troubled times was Swedish diplomat Raoul Wallenberg (1912-47); believed by some to have been the ultimate humanitarian hero of the Second World War, having saved thousands of Hungarian Jews from almost certain death. Hungary, formerly an ally of Germany in the First World War, believed that its future lay in active cooperation with Nazi Germany. In September 1942, Berlin requested that the same measures imposed on German Jews be applied to those in Hungary (including the wearing of the yellow star). However, shortly afterwards the war began to turn in the Allies favour, and Hungary began to look for ways of extricating itself from the war – incurring Hitler's wrath in the process.

As a result eleven German divisions entered Hungary, in March 1944, emboldening Hungary's very own fascist **Arrow Cross** militia; who then assisted in the wholesale identification, ghettoisation and deportation of Jews.[35] Contemporaneous with this development were the activities of Raoul Wallenberg, who helped to issue diplomatic passes to Hungarian Jews; and even succeeded for a time in establishing an international ghetto, where those possessing such papers could concentrate safely. Many of those acquiring an uncertain domicile in this manner finding a permanent home in the state of Israel, created just a few years later.

The Allied war machine was, by this time, closing in on Berlin. And, by December 1944, Germany was the sole remaining Axis power in Europe (Hungary having been effectively annexed by Germany earlier in the year). The apparatus of the Holocaust remained functioning well into the closing weeks of the war, probably due to its association with the German military. Therefore, decapitating the German military machine appears to have been the most effective way of ending the Holocaust. Together, the Allied Forces were more well-balanced than the forces of the Third Reich – none more so than the Western Powers, who quickly distanced themselves from the Soviets, following victory in Europe. As for Raoul Wallenberg, he died under suspicious circumstances whilst in Soviet hands, following the liberation of Hungary.

Like that other eminent character in our story, Sir Winston Churchill, Raoul Wallenberg was made an Honorary Citizen of the United States, albeit posthumously. Today he is remembered as one of the truly outstanding figures of the last war, and is celebrated at Israel's Yad Vashem Holocaust Memorial as one of the righteous amongst nations. The forces of the extreme Right – so patriarchal, mentally unbalanced, grossly externalizing and heavily enriched by propaganda – had been defeated. Only to Western eyes the forces of the communist Left appeared to contain many, if not all, these same deficiencies. Therefore, as the fascist threat past into history, an enemy-in-waiting raced to procure German nuclear secrets.

- ## Interfering with the destructive

Slovakian Jewish leaders first suggested to the Allies the idea of bombing the railroads leading to the Nazi death camps, and even the camps themselves, in May 1944. One explanation for the Allies reluctance to redirect air power in that way relates to the punishing effects of the atomic bomb. In chapter 2 we saw how America's Manhattan Project had committed itself to building the world's first nuclear fission bomb. By 1943 enough was known about the feasibility of such a weapon that the Western Allies were growing increasingly anxious as regards Nazi Germany's very own nuclear weapons programme. Consequently, the US established the so-called Alsos Missions, to monitor the progress of nuclear fission technology in lands held by the Nazis.

This peril was compounded, as the war drew to a close, by the Soviet advance on Berlin. Potentially, German nuclear materials and scientists might fall into Russian hands. Remember, the American's first successful detonation of a nuclear device – in New Mexico, on the 16 July 1945 (the so-called **Trinity Test**) – actually took place after Victory in Europe Day. So, the earlier race to reach Berlin, acquire nuclear weapons technology, and end the war in Europe, had proven all-consuming in those final months. Forestalling the Russians, as regards this unprecedented capability, became the single most important objective of the Western Allies at this time; second only to the unconditional capitulation of both Germany and Japan. Thus, the Nazi death camps became a low priority, as the end-game approached.

The Western Allies simply couldn't tolerate their ill-balanced opponents gaining a nuclear lead – given that all those affected in the present would be but a minute fraction of all who would subsequently be affected. Following the Trinity Test, which successfully demonstrated the ferocity of these weapons, the Americans proceeded to lay waste to the Japanese cities of Hiroshima and Nagasaki. But the question obviously arises as to whether this was a sparing use of punishment, sufficient to define civilized limits – or whether it was an excessive and unruly use of force. The received wisdom is that it was a well-balanced act which brought a punishing conflagration to a close, not just in the pacific, but globally. In its place came a rewarding postwar environment – one which the Japanese themselves have benefited from enormously.

In the war's immediate aftermath the global environment became dominated by mounting tension between Russia and the West. That enmity being cemented by the division of Europe, including Germany itself, and by the Soviet's successful testing of an atomic bomb (in August 1949). According to the writer and academic, Professor Tony Judt, "the scale of the punishment meted out to the citizens of the USSR and Eastern Europe in the decade following World War Two was monumental – and, outside the Soviet Union itself, utterly unprecedented".[36] Thus, the West was caught in an uneasy peace with an ill-balanced opponent – with both sides possessing a growing stockpile of nuclear weapons. This, then, more or less set the tone for the next 40 years.

It was now a waiting game as regards the demise of the communist Left. The Soviet bloc would eventually show signs of insipient economic collapse, leading to widespread fears in both Britain and America that the Soviets might grossly externalize their internal stresses. But America needed to watch its back, lest it become prey to the gross externalization of stresses arising within its own population. In that respect the *trias politica* may seem like a

126

palliative, i.e. a way of targeting or controlling the symptoms arising from an ill-balanced citizenry, rather than the underlying destructive forces themselves. Like the individual psyche, the careful crafting of America's character was proving frustrating, due to the unyielding nature of some of its own dysfunctional subnational elements.

Ch11: **Raising the individual**

- Playing with plasticity

The clinician studying the development of children might define **nature** as that which isn't chemically-imprinted onto the fabric of the brain as a consequence of experience, but which derives from the chemistry and fabric itself. Conversely, the manner in which exogenous factors shape the individual, through memory, learning and experience, and the chemical imprint they leave behind, may be described as **nurture**. Begging the question "was it nature or nurture?", when confronted by disturbing actions or events. The Freudian might perceive the psyche as comprising both nature *and* nurture; with the id approximating the former; and the ego and superego the latter. In theory, another's nature could be superimposed, free of the inhibitory effects of one's own nurture – hence the logic of losing nuclear secrets to the Russians, in order to consume their finest minds.

The term **plasticity**, by which we mean the malleability of a person's character throughout their development, when subject to inculcation, experience and stress, correlates with the individual's potential for nurture. A person's nature, however, isn't so pliable, unless compromised by neuropathology. Because nurture imprints itself onto the very fabric of the brain, via afferent and efferent signals, and because those signals can be mediated by the mind of some third party, that makes superimposition an aspect of nurture. Thus, the act of superimposing on a child takes nurturing to whole new levels, with some children appearing more malleable than others when subject to this deliberate re-shaping of their character.

In the future much will be made of a child's plasticity; by which we mean the ability of others to impress new ways of thinking, acting and behaving onto the medium of the child's brain. The only conceivable limits being the child's own unyielding biology. In the final analysis, art critics all agree that whatever else art is, it ought to be authentic – and so it is with personality and behaviour. But this, of course, begs the question: what *is* truly authentic and how do we provide for the same. Essentially, authenticity equates with the personality and behaviour arising when an individual is placed in a balanced environment, with this well-balanced environment providing the template for democracy. That is to say, by conforming to the best of women's instincts, NEURON would guarantee to reveal the child's authentic inner self.

Conversely, an ill-balanced environment will render the authentic inner self incomprehensible. The fascist Right and communist Left would have replaced authenticity with bogus mass conformity. Democratic constitutions provide for all things, save and except those things which are proscribed by common consent. Moreover, democracy provides for all things arising under balanced conditions; and for the limits being debated and defined under the self same circumstance. Assuming the phenomenon is suitably revised those limits would be authentically policed, and in a manner which conveys the truth, regarding the logic, wisdom and integrity of those involved. Crucially, punishment is left to the mainstream, as the capacity to deform personality and behaviour (via induced effects) is simply too great.

Democracy affords the state a great deal of plasticity, just as memory and learning render our responses elastic through enrichment. This all-encompassing plasticity, which spans the range from a given individual through to state level, is the antithesis of totalitarianism. If the state is defined as a person in international law, then the executive is the id, the legislature the ego, and the judiciary the superego. Executives are generally unyielding in their core principles; unlike the legislature and judiciary, whose complexion changes. This very plasticity is a boon to a supranational presence, as it makes exercising its powers easier and more meaningful. Ill-balanced politics and individuals will still arise, but NEURON would have a duty to correct such imbalances – either directly, through rewarding the right people; or, indirectly, by supporting and assisting punitive mainstream activities, e.g. imbuing those in the law (who excel) with certain faculties.

- **Language and perception**

Ambiguity underscores the importance of learning in language comprehension. Or, put more clearly, you are unable to perceive the intended meaning of an ambiguous sentence because it conflicts with what you've been taught. Making sense of things relies heavily on learning, and central to such learning is language. The single most important phase, as regards language development, is the first few years of life – known as the **sensitive period**. After the age of eight acquiring language skills becomes much more difficult. Complex emotions probably evolved prior to the birth of language, greatly frustrating our earliest ancestors; just as poor language skills today greatly frustrate young minds. With language evolution and acquisition, however, comes the peace of mind afforded by effective communication.

This obviously begs the question as to whether the proffered neurological organization should involve itself in the lives of those too young to accurately perceive what is happening. One might specify certain criteria which must be met (in terms of age, ability, comprehension and awareness) before neuro-cognitive effects can be visited upon the young. Having said that, superimposition takes nurturing to all new levels, and those criteria might conceivably be met rapidly whilst under its influence. Viewed in terms of **social stratification**, those with a good command of language might be said to meet the demands of telepathic involvement sooner. As for the truly vulnerable, never visiting effects at all would at least ensure that they never become hollow vessels, actioning behaviours at the behest of third parties.

These issues neatly dovetail with related concepts such as continuous and discontinuous development (**continuous development** being smooth and regular; and **discontinuous development** being irregular and episodic). In order for the child to comprehend that telepathy isn't natural its assimilation should be largely discontinuous, involving schooling in an episodic manner. Ultimately, the challenge is to ensure that the resulting youth has a manner and a presence which is authentic, rather than fake or artificial. And, that the person is able to adjust meaningfully to this new and unusual climate – possibly becoming a more resilient person in the process. **Resilience** being the ability to recover quickly from adverse effects, or the tenacity to continue in spite of negative experiences.

The fact that a person can be rendered more resilient suggests that nature isn't everything, and that the potentially unresolved might be helped to become more resilient through a process approximating nurture. Throughout the process of development various **personal constructs** undergo a process of discontinuous adaptation, assimilation and revision; sufficient for the person to arrive at an overarching view of themselves and their world. Approached correctly, a person reaches **equilibrium**, sufficient to capitalize on the experience of knowing – such that extrasensory effects can be accommodated within the limits of their present understanding. A wider **disequilibrium** would result from not being able to explain such effects within the present scientific paradigm.

The act of writing mirrors constructivist theory, in-so-far as it involves sentence generation and revision. According to many standard texts on writing, there's no such thing as a good writer, only good re-writers.[37] Similarly, many psychologists believe that the script underlying a given psyche would benefit from periodic revision. That careful re-crafting of one's character affecting the way in which the individual supplements the information received by their senses. Whilst erroneous knowledge and illogical thought processes can still derive from this constructivist process, so too can supremely well-balanced minds and global equilibrium. Evil will triumph, if good people enrich the information conveyed to their senses in an inappropriate manner. But armed with the right ideas, and a suitable command of language, progress is assured.

- Polymorphism

Coevolution, otherwise known as duel inheritance theory, argues that genetic evolution and cultural evolution are intertwined. That culture impresses itself onto the brain through social learning, thereby affecting the behaviour of the said adults. The resulting adult behaviour, which stems from cultural enrichment, has the effect of dictating the physical characteristics of the next generation; which is then shaped by its own, more modern culture – giving rise, in turn, to another generation, similarly shaped by its own biology and the culture it keeps. Thus the manner in which we enrich the information received by our senses may ultimately play-out on a grand scale through evolutionary processes. The premise behind coevolution is that mankind is significantly affecting its own evolution through the culture it chooses.

If we adopt the model of balanced-mindedness as our chosen culture (namely, favour rewarding; use punishment sparingly, and only to define the limits), and adhere to it long enough, we should begin to see evidence of the same arising in the gene pool. Eventually arriving at a society less heavily-reliant on external conditioning, self-improvement and learned behaviours. But even then, intrinsically well-balanced minds remain susceptible to misinformation; this being the Achilles heel as regards otherwise profoundly well-balanced individuals. Unless, of course, those profoundly well-balanced individuals are inducted into NEURON; in which case they become party to privileged knowledge, free of the vagaries of mainstream intriguing.

The fact that humanity differs largely in terms of appearance (i.e. colour, physiognomy and build), suggests that duel inheritance is a rather superficial process, related largely to **sexual selection**. Thus, humans are quintessentially polymorphs, possessing differing eye

131

colour, body mass, hair type, etc; with coevolution the driving force behind this **polymorphism**. My point is this, **evolution** in modern humans remains a relatively superficial process related largely to appearances, with the culture we keep placing the greatest emphasis on how we look – balanced-mindedness being a relative abstraction. This reflects the fact that vision is by far our most dominant modality, transcending even extrasensory awareness. The question now is whether we are prepared to appropriate a culture transcending vision.

Therefore, one could argue that *Homo sapiens* hasn't evolved, save and except in a superficial way. This counter-evolutionary model implies that the fabric of the brain has remained much the same, and with it our source of primitive drives and desires (the ego and superego changing, but not the id). Thus mankind's deepest, darkest and most unruly impulses continue to shape our actions, as much today as in the distant Paleolithic past. What's really changed is what we impress onto the brain, culturally. But as the culture we keep is so often facile, particularly amongst the most sexually-active, coevolution seems doomed to be a relatively superficial process. Culture has therefore served to differentiate between the races, but not to conjure-up fundamental differences. Hence endemic violence and ill-balanced feuding across all ethnic groups.

Not only have we failed to significantly evolve, we still require a robust constitutional framework to contain and control the effects of our defective genome. The *trias politica* exists to compensate for the unconscious drives and endemic contradictions present in all racial groups. It's sobering to think that any one of us could be punished on the basis of our thinking, if thinking were indeed punishable. However, one shouldn't proscribe thought, or punish cognition. Punishable behaviour is, by definition, the wrongful act, together with the thinking behind the act, never simply the thought. This concept, which is enshrined in law, would bring telepathy into line with current legal practice. So, the democratic system compensates for an endemic imbalance – one which is unlikely to be cured by coevolution, if the culture we keep is entirely superficial.

- Who's bad?

Conscience theory paints a somewhat Pavlovian picture of guilt, misconduct and correction, i.e. wrongdoing (*conditioned stimulus*) meets with punishment (*unconditioned stimulus*), which produces discomfort and anxiety (*unconditioned response*), resulting in the avoidance of misconduct (*conditioned conscience*). One might term this corrective imposition, **heteronomous morality** (which is a morality defined principally in terms of consequences and externally-imposed controls). Heteronomous morality is commonly associated with the correction of young children and the defining of strict legal limits. Taken to an extreme it results in agentic conformity and learned helplessness, of a kind commonly encountered in rogue states.

Conscience theory doesn't settle the question of what's good or bad, it simply implies that a conscience, of sorts, can be imprinted onto the fabric of a person's brain; with the efference copy militating against re-offending. Until, that is, the conditioned conscience goes extinct (in those previously too young to question, or people too weak to rebel). Heteronomous morality only emerges in adults when they are confronted with forces so powerful they can't be argued with. Allowing the phenomenon (or its replacement) to punish, leads inexorably to uncharacteristic behaviours, criminal excess, and gross violations on the part of those applying the effects – but without the question of what is truly right or wrong having been publicly decided.

Questions of right and wrong are best deliberated over in the mainstream, where punishment can be imposed via due process. In this way, a much clearer understanding of people, populations and cultures emerges. Some of those prevailing socially, politically and economically may have previously met with an *unconditioned stimulus* (or punishment); only for the mainstream to subsequently revise its limits, becoming less punishing. The impetus for such change often stems from a form of morality most commonly encountered in adults across the free world – namely, **autonomous morality**. Autonomous morality is a reflection of character, rather than consequences. And, if enough people think that something isn't sufficiently deviant as to warrant punishment, democracy will force a liberalization of social norms.

So, what is bad? Well, if truly well-balanced people define reasonable limits with the sparing use of punishment, revising those limits periodically so as to be as reasonable as possible, then bad is transgressing those most reasonable of limits. But this presupposes that those applying neuro-cognitive effects aren't ill-balanced, and that their powers haven't been in some way misappropriated. If those powers are being misused, good people will appear disturbed, dysfunctional and rebellious; while bad people will appear agentic, self-interested and acquiescent. Even

133

when such effects are conscientiously applied, one is still left with a mainstream debate as to where the absolute limits should lie. Therefore, it isn't possible to be definitive as regards what the limits should be without appearing dictatorial.

In many respects the *conditioned stimulus* (or wrongdoing) is synonymous with injurious, destructive and ill-balanced forms of behaviour. Which, on a grand scale, equates with the illegitimate use of power at the national level; such illegitimacy forcing the hand of well-balanced people in the direction of conflict, conflagration and violence. Twice in the 20th century the world was engulfed in total war, prompting the development of ever more terrifying weapons – weapons capable of concentrating in the wrong hands, making war more terrible. More than ever we require a global commitment to the legitimate use of power, and the power to prevent its illegitimate use. The model presented in this book favours states developing their own autonomous morality; with externally-imposed heteronomous morality being used sparingly, and only to define civilized limits.

•   Young people and legitimate power

Being perceived as a legitimate power by youth is crucial if you are to command their full respect and cooperation. Being legitimate means conforming to the aforementioned axiom or rule. Those limits, referred to in this rule, being subject to democratic debate and periodic revision. What's fundamentally important is that power concentrates in the right hands; not that authority figures force compliance in respect of every agreed limit. By not forcing compliance in an excessive way, one can build-up a more accurate picture of autonomous morality. Character, unlike a conditioned conscience, won't simply go extinct, and can be relied upon in times of crisis. Judging character and concentrating power wisely are more important than zero tolerance. And, never more so than in the postmodern age.

In this postmodern era, with all those erroneous grandstanding political ideologies behind us, the priority is to avoid power being appropriated by the wrong people – precisely the kind of people who are more likely than most to engage in zero tolerance in respect of some rather arbitrary limits. The western polyarchies should remain true to liberal individualism – and, by implication, personal autonomy. But with that commitment comes the danger of unruly politics arising out of the stresses of those all-pervading political, economic and military climates. To counter the same we require individuals with sufficient character that they might fairly be rewarded with extrasensory abilities. Some of these more deserving individuals being inducted into an overarching scientific and technological presence later.

In fact, it may be easier to deduce who these young people are when they are confronted by political pathology – although their motives for taking on unruly others may be mixed, e.g. financial gain, self-preservation and a sense of camaraderie. It can't be overstressed that the biggest danger we face is unprecedented powers falling into the wrong hands, and that democracy, autonomous morality and personal freedom are allies when it comes to judging who the right people are. However, the heavy-handed application of neuro-cognitive effects will undermine democracy, force heteronomous morality and deny people the freedom to be themselves. Within such an environment all power risks concentrating in the wrong hands. Mainstream authenticity, however highly-charged, does at least allow us to perceive accurately.

To avoid the obvious limitations of **classical conditioning**, in which conscience is little more than a trained reflex, we need to provide for the proper development of children (within a conceptual framework which does justice to the complexities and contradictions self-evident within the human psyche). An environment in which youngsters can think with complete impunity, becoming authentic personalities in the process. And, although autonomous youthful rebellion may doom your attempts at defining everyday limits, take comfort in the thought that you are a legitimate power. One day they'll be old enough to be presided over by other far more expansive legitimate powers, equally committed to being, first and foremost, legitimate.

- Domains

Down at the national and subnational level we encounter various domains, categorized by **social domain theory** as: 1. the moral domain (which includes the most serious offences); 2. social-conventional domain (which includes breaches of local traditions and values); and, 3. the personal domain (involving minor contraventions of non-regulated norms). Well-balanced individuals aren't going to labour over infractions at levels two and three, but wouldn't be balanced-minded if they wholly ignored transgressions at level one. This implies that well-balanced individuals aren't going to split hairs over minor differences pertaining to local traditions and cultural norms, provided these don't conflict with a person's fundamental rights and freedoms. Thus, balanced-mindedness embraces multiculturalism and the preservation of diversity; not to mention a toleration of sardonic self-mockery and satirical irreverence. Religious toleration is the norm, as is the expectation that those who are religious will be at least as tolerant as their more secular counterparts.

This then, is the free world, from which all those wielding neuro-cognitive powers are to be inducted; and where to be inducted you must reflect such freedoms. Having provided for autonomy and freedom of expression, individuals are expected to manage their own behaviour – provided that the chemistry of one's soul isn't being compromised by heavy-handed interference. One particular aspect of social psychology is effortful control, self-regulation and restraint, which together denote the ability of an individual to control their own thoughts, feelings, actions and behaviours. Undermining inhibitory responses needn't be as dramatic as intermanual conflict, it might simply mean saying something which was best left unsaid, doing something which was best avoided, or, in the case of impressionable youth, being that bit more impressionable.

Adolescents are inclined to convey their emotions freely (but not always constructively), and so it's conceivable that unruly forms of interference might force the issue in the direction of intra-family conflict. A male-dominated phenomenon, impatient for stimulation, may be more inclined to generate such conflict than one with a revised demography. So, the suspension of inhibitory signals could negatively impact on the behaviour of youth, resulting in behaviour which might otherwise be questioned or avoided by the youth themselves, e.g. binge-drinking, illegal drug-taking, unprotected sex, and even risk-taking (with the potential for death and serious injury). The much needed paradigm-shift would relieve youth of the most serious violations – such excesses having become inexcusable infractions under present circumstances.

Postwar America witnessed the rolling-out of radiologically-induced effects, sufficient to invigorate the whole of its culture. Unfortunately, it may have gone further than revealing something of its subjects, it may have echoed the whims, caprices and unconscious inclinations of its male operatives. True, these functionaries have remained nameless and anonymous, whilst those under their spell occasionally burnt-out in a most public fashion – the person's self-schema (manufactured in-part by the media) enriching that person's perception of themselves, and unbalancing them in an often self-destructive manner in the process. Of course, those creating these tension may cite the rewards of eternal recognition and celebrity – which is all very well, provided that the person is an authentic presence, and not an unwitting victim of excess.

Ch12: **Personal identity**

•   Big Brother is watching you

As Richard Feynman forged ahead with presenting his ideas at the Shelter Island, Pocono and Oldstone Conferences, held between 1947 and 1949, the writer, George Orwell (real name, Eric Arthur Blair, 1903-50), was anticipating the publication of his new book, entitled "Nineteen Eighty-Four".[38] In this novel, Orwell's principal character, Winston Smith, is watched over by "Big Brother", a dictatorial power which distorts the truth, forbids privacy, outlaws independent thinking, and proscribes sexual pleasure. Much is made of audience effects, in which humiliation, self-consciousness and fear are the principal feelings of those caught-up in this dark politic. The numerous references to great purges, show trials and summary executions, leaves the reader in no doubt that the politic is wholly totalitarian.

Nineteen Eighty-Four seeks to empathize with an individual living in the shadow of an unruly one-party system, possessing unrivalled powers; and which insists on bearing down on the proletarian majority beneath. In many respects, Orwell was anticipating the postmodern era, with his active distain for Marxist-Leninist style subservience. Language acquisition and a decent education, which together would empower the masses, are actively controlled by "the Party", in order to make "thoughtcrime" ultimately impossible. One perceives Winston as living and working within a complex web of interconnectedness, and suffering as a result of that involvement. Far-fetched though it may seem, it nevertheless predicts Pol Pot's murder machine, with its disconcerting take on human nature.

Big Brother is the manifestation of supreme power, concentrated in too few hands, and to such an extent that it finds itself gratuitously externalized – particularly in respect of its own people. In many ways an illegitimate power, such as that conjured-up by Orwell, is simply illegitimate; i.e. one needn't invoke surveillance or mind control to decry what Big Brother stands for. Students of Orwell's book are faced with a dilemma – are they disturbed by the thought of surveillance and mind control, like notorious US National Security Agency whistleblower Edward Snowden (1983-), or by the illicit use of power? Could you, for example, conceive of a politic which actually provides for surveillance and mind control, having determined that such powers might be usefully employed. Science and technology are humanity's way of explaining and exploiting natural forces – but far from being the slaves of politics, as in

Orwell's novel, neuro-cognitive phenomena would be distinct from the same in the model spoken of in this book.

Ultimately, Nineteen Eighty-Four is about an illegitimate power, which happens to possess certain abilities; but that shouldn't deter us from arriving at a legitimate power, possessing a similar potential. Provided punishment is confined to the mainstream, and the most scrutinized are those preaching surveillance at the top, then maybe a suitable politic can be arrived at, one which strengthens the hand of the free-world, rather than compromising the same. The fate of the individual within such an environment fascinates us all, hence our curiosity as regards Winston Smith, the fictional inhabitant of London, chief city of Airstrip One, Oceania. Orwell wrote, "How could you make an appeal to the future when not a trace of you, not even an anonymous word scribbled on a piece of paper, could physically survive". Which sums up the punishing impact of totalitarianism, and just how dismal the rewards must appear to those caught-up in the same.

The Marxist-Leninist nightmare vision, presented by Orwell, ironically gave credence to growing anti-communist sentiment in America. This was, after all, the start of the Cold War, not the endgame, and one senses a certain amount of anxiety as regards the promulgation of purported scientific fact. Crucially, Feynman had to win every single argument in order to create the smokescreen that we've lived with ever since. A photograph of Feynman talking to a peer (particle physicist Murray Gell-Mann), dated around that time, betrays a good-humoured unease on the part of the former. The whole of the West's Cold War strategy rested on those in-the-know talking-up particle physics, whilst playing-down the significance of fields – sufficient to disguise the neuro-cognitive power coming into play.

That history played-out under the gaze of a phenomenon which was nominally supranational; transcending both the national and subnational politics of the country which had spawned it. Already well beyond the reach of irascible, highly-misogynistic and overtly-racist subnational elements, it was becoming an invisible hand acting mysteriously in pursuit of long-term objectives. Some of those who would subsequently meet their end under its gaze were progressive-minded individuals, sacrificed in pursuit of those goals. If not entirely misogynistic itself, the phenomenon was, nonetheless, inherently patriarchal. As for its chief political goal – presiding over the demise of fascism and communism – that was still only half-finished by the time of Orwell's book.

The West was still piecing together a whole new orthodoxy in physics at this time; with the giants of political and military intrigue being less than entirely honest with their Russian counterparts. Elementary particles took centre stage as regards our broader understanding of mass, charge and energy; with the resulting blind-spot in popular scientific knowledge being rendered impenetrable through the talking-up of various assumptions. Within this cloak and dagger atmosphere, George Orwell seems like a sensitive plate exposed to invisible rays, tormented like his contemporary, Virginia Woolf; sufficient to reduce him to walking the streets of London and Paris. He would have perceived those applying the effects as an illegitimate power. But, to some extent all who experience such effects interpret them as illegitimate, such is the nature of that initial fatalism. This is, in fact, an unavoidable paradox, resulting from the testing of the human spirit.

What is clear, is that Orwell externalized his issues responsibly, via the written word, and in the manner of a person possessing a well-balanced persona. His was no immature mind harbouring a misplaced autonomous morality. George Orwell was a person saying to themselves "without victory, there is no survival".

- **The Vietnam War**

In the postwar era, to be anti-communist was to be pro-American, irrespective of whether you were British, French or Vietnamese. Cleaving loyally to America was perceived by most developed nations as being preferable to being non-aligned or neutral, especially following the Suez Crisis of 1956, in which Britain, France and Israel's offensive against Egypt (to secure the free-movement of oil) went disastrously awry. Thereafter, the world divided along largely pro-communist, pro-American and non-aligned lines. Vietnam, by this time, had been formally divided along the 17th parallel, due to irreconcilable differences between the communist-affiliated North and the pro-American south. As the Cold war intensified, Vietnam became an obvious flashpoint for potential future conflict due to ongoing enmity between the Soviet backed north and American supported south. Two rifle bullets was all it then took to engulf Indo-China in the bitterest of conflicts.

The victim of those two rounds was America's youngest and most popular President, John F Kennedy (1917-63), otherwise known as JFK.[39] The year prior to his assassination his alleged lover, screen goddess Marilyn Monroe (1926-62), grossly internalized her troubles, resulting in her suicide. The evidence at that time suggests she was unresolved, conflicted and suicidal – with her death an inevitable consequence of those three factors. JFK's male assassin, Lee Harvey Oswald (1939-63), on the other hand, had no qualms about grossly externalizing his inner pain. Precisely what his issues were was never determined, as Jack Ruby (1911-67), a nightclub owner, externalized his anger at the assassination by shooting Oswald dead at point blank range.

As for the phenomenon, it may have stood down, allowing Oswald to fire those two fateful shots, believing that JFK's extrajudicial attempts at removing communist Cuban leader, Fidel Castro (1926-2016), were either too dangerous, too unlawful, or simply too unnecessary (given the neuro-cognitive capability coming into play at that time). Besides, America had made a huge financial commitment to putting a man on the moon, and the cost of even a limited conventional war on the ground would have been compromising enough of that aspiration, without inviting thermo-nuclear war. Lyndon B Johnson (1908-73), the US Vice-President, immediately assumed the role of President; being formally elected to that office in 1964 with a huge majority.

In that same year, the American Congress passed the Gulf of Tonkin Resolution, in response to an alleged attack, against US destroyers, by North Vietnamese forces. As a pretext to war the Gulf of Tonkin incident was *very* low-key, not at all like the 9/11 incident, 37 years later. It was beginning to look as though the actions of one disaffected American male had led to a sea-change in foreign policy in Indo-China; one which would result in the eventual deaths of many, often disaffected, American males – during what became known as the Vietnam War (1964-1975).

As this 100 billion dollar war in Vietnam escalated, tensions began to mount, with the information conveyed to people's senses around the world becoming ever more enriched by the media. Months of fear, frustration and anger resulted in the My Lai massacre, in the Quang Ngai province (Mar 1968), as disaffected American troops engaged in indiscriminate killing and torture. Back in the US, civil rights leader Martin Luther King Jr (1929-68) and the former President's brother, Robert Kennedy (1925-68), were similarly gunned down by mentally unbalanced male assassins. The overall impression is one of nature and nurture being heavily supplemented by exogenous forces, sufficient to lead to a violent externalization of real issues.

As the cost of the Vietnam War began to mount, its effects began to be felt globally – hinting at the ruinous implications of a direct nuclear exchange between east and west. In 1971 the US President, Richard Nixon (1913-94), ushered in a whole new era of floating exchange rates. This floating of the American dollar (in part, a reaction to the Vietnam War) had an adverse effect on many European economies; with its effects further compounded by the first of two oil shocks (i.e. the one mentioned in chapter 9). As the Vietnam War came to a close, with American capitulation and the fall of Saigon, monetary inflation took off in the **UK** (exceeding 24% in 1975). To the more observant, it was starting to appear as though Lee Harvey Oswald had destructively interfered with the UK economy.

In Britain, rampant inflation, growing currency speculation and rising import costs resulted in a winter of discontent (1978-79), together with rising unemployment and a bail-out from the IMF. The principal beneficiary of this widespread despair was Margaret Thatcher (and the Conservative Party), who began to gain popular support in the UK. By this time all persons of note were capable of being known remotely, via radiofrequency technologies. Some of these individuals, be they politicians, diplomats, actors, rock stars or assassins, might even have interpreted their actions as being entirely voluntary. In the final analysis, destructive though the conflict in Vietnam was, it avoided far more punishing alternatives – ones capable of killing in one hour the numbers lost over the course of a decade.

So, what happened to all that money saved through conventional skirmishes and much publicized posturing? Well, in the case of America it funded new heights of scientific and technological progress; with NASA's projects Mercury, Gemini and Apollo (1958-72) culminating in the moon landings. It seemed as though humankind was facing a straight choice between investing in progress or the unruly alternatives. Those applying neuro-cognitive effects have always appeared keen to exaggerate the underlying chemistry – and even, on occasions, augment it. And so it is in our collective interests to think very carefully about the kinds of chemistry we indulge. We are, after all, whatever that chemistry happens to give rise to. Besides, the person unbalanced by unruly interference might one day succeed in altering the course of history.

•   **Protect and survive**

During the early years of Margaret Thatcher's premiership (she having been British Prime Minister between 1979 and 1990) the world was gripped by a 'Second Cold War'. The Soviet invasion of Afghanistan, in 1979, and their siting of SS-20 medium-range nuclear missiles in the Ukraine (countered with Pershing-II missiles in Western Europe), ended the constructive détente which had previously existed between east and west during the 1970s. This overt souring of relations was made more ominous by two things: firstly, the Americans and Soviets had stockpiled so many nuclear warheads that people now talked in terms of **mutually assured destruction** (MAD), and secondly, the communist economies in the east were falling into serious disrepair. Ironically, had the Russian people had more influence over their choice of leader things may have been even more dangerous.

Contemporaneous with this right-of-centre Conservative Party victory in Great Britain was the election, in America, of the Right-wing Republican President, Ronald Reagan (who held that office between 1981 and 1989). Anglo-American relations were particularly good during these years, further strengthening the NATO alliance, which was so desperately needed. However, neither premier was entirely immune to the effects of remote interference and later both Reagan's and Thatcher's minds would, like the Soviet economy itself, begin to show signs of serious disrepair.

Back in the initial years of their friendship, deteriorating relations with the Soviet Union had made Western Europe uneasy – hence the 'Protect and Survive' public information campaign which ran in the British Isles during that time, and which sought to educate people, as far as possible, as to the dangers associated with fall-out and nuclear attack. Some might even interpret the **Falkland's War** (1982) as having been a heaven-sent opportunity to bloody the proverbial troops, prior to a ground war in Europe. Had the public fully realized the state of insipient collapse emerging in the east they might have been more aware of the frightening possibilities. As it was, they were far too caught-up in their own economic woes at home to do much more than satirize those concerns.

The anti-inflationary priorities of the Thatcher government led to rising unemployment; which was compounded by the collapse of manufacturing and the loss of unprecedented numbers of mining jobs. Within this climate of comparative gloom and doom the sum of all international bank lending grew from $324 billion a year (1980), to $7.5 trillion (1991); a 2,000% increase.[40] One can only speculate how much of this borrowing went towards funding radiofrequency effects in the run-up to Communism's final collapse. But, given the Conservative Party's commitment to selling-off nationally-owned assets, particularly state monopolies, the British government was obviously less keen than others when it came to creating holes in its generational accounts.[41]

The received wisdom regarding the final collapse of communism in the east, and the subsequent disintegration of the USSR, begins in Poland in 1980. Here price rises brought mass-backing for the unofficial trade union 'Solidarity', which became the first officially recognized trade union in a communist country. This softening of the party-line paved the way for Mikhail Sergeyevich Gorbachev (1931-) becoming Secretary General of the Communist Party of the Soviet Union, a position he held from 1985 until 1991. Widely perceived as a reformer, he became associated with terms such as *glasnost* (openness) and *perestroika* (reform). His immediate concern on taking office wasn't the West, but rather his own economy.

Amid the widespread political and economic reforms at home, Gorbachev capitulated in respect of Afghanistan, signing a United Nations initiative which pressed for the withdrawal of Soviet troops from the country. The Russians subsequently withdrew in 1989, partly due to their own domestic problems and partly due to the ungovernable nature of Afghanistan itself. Mujahedin rebels, including al-Qaeda, fought for control of the country, sweeping away all signs of progress in the process. Out of this civil war emerged an Islamic revolutionary movement, termed the **Taliban**, which assumed control of the country. Under its strictures women (who were now obliged to wear the **burqa**) were forbidden to work outside the home or receive an education. The Russians, who'd been suppressing or at least drawing the fire of al-Qaeda, were gone. Al-Qaeda, and Osama bin Laden in particular, were now looking for new opponents beyond the confines of their dusty rebel strongholds.

Back in the Soviet Union, having eviscerated the Communist Party dictatorship from within, Gorbachev presided over the USSR's peaceful fragmentation. On November 9th 1989 the East German authorities announced the opening of the Berlin Wall. Within one year of this momentous event the 'Final Settlement with Respect to Germany' was signed in Moscow, formally reuniting the two Germanys. With Germany officially reunited, Gorbachev then abolished the USSR, and with it his position as Secretary General – both of which ceased to exist as of midnight, December 31st 1991. The following day the world woke-up to a whole series of independent republics, together with a **Russian Federation**, commanded by Boris Yeltsin, who'd resigned his membership of the Communist Party (see Fig.18).

Figure 18: Borders of the former USSR and the present Russian Federation

Russian Federation

Former limits of the USSR
(other members of Warsaw Pact not shown)

Whilst many republics had seceded from the USSR with comparative ease, others found withdrawing from the Russian Federation, which replaced it, impossible. One of these regions was the Republic of Chechnya, which strove for a time to assert its independence through the actions of Chechen Muslim extremists. One of the most disturbing acts they committed was the storming of a school in Beslan, North Ossetia. The Beslan School Hostage Crisis (2004), as it became known, resulted in more than 1000 people being held hostage, in an act that many would interpret as deranged. By its conclusion, three days later, more than 300 hostages were dead – over half of them children.[42]

Terrorism is a symptom of political and military weakness. Horrifying though those events were, real power lay in the hands of more adequate others. Provided that power concentrates in the right hands there will always be those unruly elements who attempt to influence from a position of weakness, and in ways which are wholly ill-balanced. Thankfully, for those on the continent, those deranged few weren't in Moscow presiding over the fate of millions. Had they been, Europe would have been ablaze, with countless numbers dead within its embers.

- ## Pan-Arab struggle

With the dissolution of the USSR the phenomenon's principal reason for existing had vanished – the political extremes of Right and Left having been decisively confined to history. Now the two greatest hazards facing the free-world were, firstly, the phenomenon itself (which was drifting without a clear remit), and secondly, the growing number of anti-Western Jihadist movements, of which al-Qaeda was but one, appearing worldwide. Reforming the phenomenon was the priority, so that whatever subsequently happened couldn't be attributed to its own unruly practices. Provided that power is legitimately applied at the subnational, national and supranational levels, then unruly behaviour on the periphery of mainstream politics becomes indicative of the pathologically-minded being weak.

Figure 19: Middle East (after World War I)

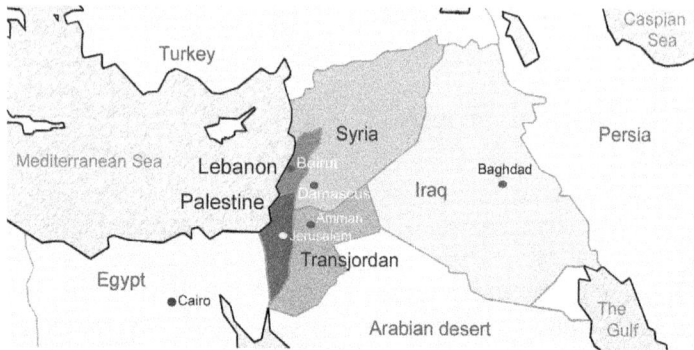

This debate has particular significance in the Middle East, where Western influence is often perceived as illegitimate, usually by ill-balanced elements inclined to terrorism. Current Western involvement in the region dates back to the end of the First World War, when France gained control of Syria and Lebanon; and Britain got Palestine, Transjordan and the newly-formed state of Iraq (see Fig. 19). As for the kingdom of Saudi Arabia, that formed in 1932, following a civil war in the region. Ironically, for both the French and British managing the thorny politics of their new acquisitions, American prospectors quickly discovered considerable quantities of oil in the new Saudi kingdom, leading to rising exports of the same. Thus, Saudi wealth grew, as the British faced a long-winded and highly fractious debate regarding the future of Palestine.

Zionists had long argued for the creation of a permanent Jewish homeland in Palestine, and repeatedly lobbied the British to that effect. The resultant **Balfour Declaration** (1917) and **White Paper of 1922** supported the establishment of a national home for Jewish people in Palestine, but stressed that the solution should neither alarm Palestinian Arabs nor disappoint the Jews. By 1939 the British appeared to have distanced themselves from the Zionist cause, causing anger amongst those who feared it would be a death sentence to European Jewry. Their fears were well-founded, given the nature of German **anti-Semitism** at that time and the holocaust which followed. In 1948, the British finally surrendered control of Palestine, enabling the future Israeli Prime Minister, David Ben-Gurion (1886-1973), to establish the State of Israel by force in their absence. With the support of the United States, Israel successfully weathered the storm surrounding its creation.

Some might argue that the Arabs were nominally allied to the Axis Powers due to the activities of Arab nationalists such as Haj Amin al-Husseini (better known as Haj Amin). Haj Amin rose to political prominence as a vociferous anti-Zionist, committed to the preservation of Palestine as an independent Arab state. He played an active part in the Arab Revolt in Palestine in the late 1930s, and courted the anti-Semitic Third Reich as a means of advancing the Arab cause.[43] According to this argument, the Allies (USA, Britain, Soviet Union, etc) who defeated the Axis Powers, had as much right to preside over the future of the Middle East as over Germany itself. Had they taken that initiative back then a clearer picture may have emerged.

Figure 20: Jewish settlements in Palestine and the present state of Israel

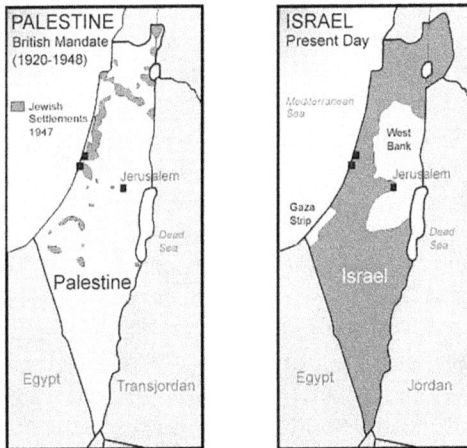

The problem was that The Allies included the Russians, who were preparing to explode their first atomic bomb at the time of Israel's creation. As Israel consolidated its position, an uneasy peace settled over the region. Then, in 1957, US president Dwight D Eisenhower (1890-1969) issued his famous ultimatum, stating unequivocally that in the event of communist aggression in the Middle East the US wouldn't hesitate to use armed force. From that point on many believed that nuclear conflict would originate in the Middle East – and that concern never fully evaporated until the formal disintegration of the Soviet Union, 32 years later. Today, Arab nationalists are faced with Western Powers who have stared down and defeated both fascism and communism in their most militarized forms. At best, therefore, they might achieve reward and recognition through democratization, liberalization and wholesale legitimacy.

Since the **Eisenhower Doctrine**, Israel has expanded its size by a factor of four (mainly due to the 6 Day War, of 1967), displacing many Palestinian Arabs in the process (see Fig. 20). Between them, however, Palestinian nationalists, the United nations and International pressure have done much to contain subsequent Israeli expansion and hegemony in the region. The Middle East was and remains an experiment in power, i.e. its concentration, separation, fragmentation and denial. The West would rather that power concentrates in the hands of those states possessing modern constitutions; which implies a *trias politica* arrangement. That would make Israel, with its parliamentary democracy and **universal suffrage**, deserving of both political and military superiority. Conversely, the West would prefer that one-party patriarchal states, with punitive-minded cultures, remain weak. Even though such weakness often begets terrorism.

A paradigm-shift, publicly acknowledging the extrasensory capability available to the Western Powers, would underscore the West's legitimate hold on power. But with that telepathic prowess comes a promise; a guarantee that those meeting the criteria for knowing will be fairly rewarded. A key criteria for knowing being sexual equality, due to the sheer weight of available evidence, be it psychological, sociological or historical, demonstrating that women are, on average, more well-balanced than their male-counterparts, and crucially less likely to grossly externalize their inner conflicts. If the lessons regarding sexual equality aren't learnt, the consequences are likely to impact on the weak more than most. If you don't want to be rendered weak, learn the lessons.

• Endless ancillary operations

Much has been made of the so-called **Tito-Stalin split**, which occurred in 1948; whereby Yugoslavia, under President Josip Broz Tito (1892-1980), distanced itself from the Politburo in Moscow.

Communist Yugoslavia then pursued, to all intent and purposes, a counter-revolutionary model, at odds with Stalin's. The importance of this split can't be overstated. The **Balkans**, as Yugoslavia was commonly known, was perceived as the probable catalyst for wider conflict, and this split had the effect of isolating the region. That split, which was remotely engineered, had the effect of making the world a much safer place. As **Titoism** replaced **Stalinism**, one would be forgiven for thinking that the Soviet Union and Yugoslavia had become distinct entities. Yet when the USSR dissolved 43 years later, Yugoslavia instantly died. Suggesting the rift had all but healed.

Five wars over the course of the 1990s then decided the region's future. The first two wars witnessed the foremost belligerent, Serbia, being repelled, first by Slovenia, and then by Croatia. Next came a third war (a Bosnian civil war) in which Bosnian Croats and Bosnian Muslims fought Bosnian Serbs (the Bosnian Serbs besieging the city of Sarajevo and massacring Srebrenicans, in the worst example of its kind in Europe since the Second World War). Fourthly, there was a war in which Bosnian Muslims and Bosnian Croats turned on one another. And, finally, the chief belligerent, Serbia, who'd assisted Bosnian Serbs to commit major atrocities, then sought to expel ethnic Albanians from Kosovo, only to be destroyed themselves by NATO action.

The actions of the United Nations' peacekeeping force drew much criticism in those years, appearing on many occasions weak and pusillanimous. In the case of the massacre at Srebrenica an estimated 8,000 Muslim males, many only boys, were unceremoniously massacred, in an area that was supposed to be safe. Those few peacekeepers who were on hand were relying solely on reference power, i.e. power based on the fact that they had the backing and support of the United Nations. Clearly it wasn't enough. NATO action seemed altogether more decisive, taking the war to Serbia itself; sufficient to define civilized limits and bring the conflict to a close. What followed was the tracking-down of Serbian war criminals, so that they could be brought before **The Hague** tribunal on counts of **genocide**.

Fortunately, Titoism had held the country together long enough for it to be no more than a localized conflict, arising in the wake of communism. Consequently, this local conflagration failed to set the world ablaze because there was only one power bloc – the West, which confined itself to limited military action and endless diplomacy. Moreover, an overarching power arranged it that way by distancing the two communist countries, at least during Stalin's life. This may be the single most important action taken by the phenomenon at the close of the Shelter Island Conference, adding significantly to overall global peace and security. For example, the **Trieste Accords**, brokered by the Americans and British in 1954, ended a disagreement between Italy and Yugoslavia, which could, in theory, have spiralled into nuclear war – but only if Tito and Stalin had been close.

By heart-wrenching irony, the international community's designation of so-called safe-areas (in the former Yugoslavia, in 1994, in places like Sarajevo, Tuzla, Srebrenica, Zepa, Gorazde and Bihac) was contemporaneous with the genocide in Rwanda. Even designating an area as safe became controversial, if it failed to deliver even a modicum of protection. No areas were afforded the same status in Rwanda. Although UNAMIR Force Commander, General Roméo Antonius Dallaire (1946-), now retired, did courageously organize areas of 'safe control' around Rwanda's capital, Kigali (saving an estimated 32,000 lives as a result).[44] But the viability of such areas remains an open-question, especially if the political and military will to defend them is lacking.

However, no genocide has ever gone wholly unchallenged; whether through the actions of superpowers, local militia, partisans, or individual acts of heroism. One might make a special commitment to empowering those who confront genocide; providing them with special capabilities and insights, leaving them richer in spirit for taking on that task, without actually bereaving them of the responsibility. The mainstream is a complex milieu which tests, stretches and challenges individuals; shaping as well as betraying personal identity in the process. For personal identity to be entirely authentic, and hence fairly rewarded (or condemned) power needs to be judiciously applied. With that proviso we need to define civilized limits, or be thought destructive ourselves.

According to former US Secretary of Defence, James Schlesinger, the single most important lesson of the Vietnam War was this, "go for your opponent's heart – don't be drawn into endless ancillary operations".[45] Close on one million Tutsis may have been alive today if the international community had struck decisively at the heart of Hutu Power in Rwanda. Instead, the United Nations engaged in some ancillary operations. Remember, the Allied forces brought the Holocaust to a close by striking at the heart of the Nazi war machine. To them, bombing the railroads leading into Auschwitz would have seemed like an ancillary operation.

- ## Weapons of mass destruction

By the beginning of the Cold War, Britain and France were in the process of withdrawing from the Middle East; leaving Palestine, Jordan (formerly Transjordan), Iraq, Syria and Lebanon to their respective fates. Palestine, as we've seen, became easy prey to Zionist militia, who moved quickly to establish the State of Israel on the banks of the Mediterranean (displacing many Palestinian Arabs in the process). The United States swiftly recognized the new Israeli state, believing that any displaced Arabs would be absorbed by neighbouring Jordan. Unlike Israel, which possessed a modern constitutional structure, the Arab States persisted with political systems based largely on monarchies and one party systems – rather than converting to parliamentary democracies with universal suffrage. In other words, they continued to be the kinds of states which the West would prefer to remain weak.

Whereas the western polyarchies became synonymous with the orderly transfer of power, these Arab states became associated with coups, tyranny and oppression. So it came as no surprise when, in 1979, a coup established Saddam Hussein (1937-2006) as the President of Iraq. Under his leadership, the Ba'ath Party brutally suppressed all opposition and engaged in the widespread use of chemical weapons. His invasion of neighbouring Kuwait, which Iraq justified on the basis of historical claims, led to the first of two Gulf wars. Iraq was roundly defeated in the first of these wars by a coalition of Western and Arab forces. By the time of 9/11, ten years later, Iraq was still unfinished business in the eyes of America and the UN (which would explain America's provocative military presence in Saudi Arabia; making Iraq indirectly responsible for the 9/11 attacks).[46]

Sanctions, amounting to an embargo on both arms and trade, which had been imposed on Iraq by the UN, following the Iraqi invasion of Kuwait, remained in place up until the Second Gulf War in 2003. This second conflict, which was instigated by America, and backed by Great Britain, went ahead without the full approval of the UN. The central justification for this full-scale invasion of Iraq was the belief that Saddam Hussein possessed weapons of mass destruction (WMDs). Whatever the arguments regarding WMDs, there is no doubt that the ruling Ba'ath Party had committed serious crimes against its own nationals, and that the trade embargo was, to all intent and purposes, punishing innocents rather than guilty parties.

However, the central premise of the whole invasion was that powerful weapons technology shouldn't concentrate in the hands of ill-balanced one-party dictatorships. Following the invasion, Saddam Hussein was caught, put on trial, and sentenced to death by an Iraqi Court for crimes against humanity, following a year long trial. Because Iraq, like Saudi Arabia and Iran, possesses some of world's largest oil reserves, this obviously bodes well in terms of the overall standard of living the Iraqi people can expect in the future. This particular exercise in nation-building on the part of the American-led coalition has sought to create a truly representative, multi-party democracy, possessing universal suffrage. The hope is that Iraq, in spite of domestic insurgency, will become a model which other states in the region might follow.

Iraq is overwhelming Muslim, as is every state in the Greater Middle East, with the exception of Israel (which is primarily Jewish). In the postwar era, many of the countries comprising the Greater Middle East, e.g. Iran, Iraq, Syria, Lebanon, Afghanistan and Libya, took on the mantle of rogue, failed or pariah states. But most states render themselves ungovernable, unruly or suspect without the apparatus of modern government. The Iraqi population comprises mostly **Sunni Muslims** and **Shi'ites**, who could, in theory, forge the world's most successful and progressive Muslim nation; but to do so, they must manage power wisely. Managing power wisely is as much about the orderly transfer of power as it is about its prudent application. After all, the more prudently power *is* managed, the more power the state can be trusted to possess.

This obviously begs the question as to whether the Islamic world can be trusted to possess nuclear programs, and the boundless energy (not to mention nuclear weaponry) provided by the same. Ever since the Iranian Revolution, in 1979, the West has faced the prospect of Iran's nuclear program – ostensibly used to provide nuclear power – being put to more malign use. If this highly-conservative Islamic state, watched over by a Supreme Leader (called an Ayatollah), does successfully test an atomic bomb, it won't simply join the five recognized nuclear states with permanent seats on the **UN Security Council** (i.e. America, Britain, France, Russia and China). What will happen is that it will be deemed a rogue state, with a pariah status; placing it on a par with North Korea.

If Iran strengthens its hand with a nuclear weapons capability, that might encourage other Muslim states in the Greater Middle East to copy its example (the Soviet Union aided the proliferation of nuclear weapons and so too would Iran). Moreover, it would seriously alter the balance of power in the region, setting in stone the progress which has been made. Making remote manipulation, mental telepathy and the superimposition of persons public knowledge, before Iran is in possession of nuclear weapons technology, would put the world on notice where *real* power lies. It lies in the hands of the well-balanced Western Powers, committed to advancing the cause of freedom worldwide. Should you remain unconvinced as to the credentials of the NATO alliance – and Britain and America in particular – in terms of them acting as chief arbiters on these issues, we'll examine more closely their respective politics, history and law.

PART V

Politics and Law

Chapters 13-15

Ch13: **Well-balanced politics**

• The Westminster Model

Multi-party politics began in Great Britain as a late 17th century disagreement over the accession to the English throne. Inside the English parliament two distinct factions arose over the question of who should succeed the reigning protestant monarch, Charles II (1630-85). Although Charles had fathered many children, all were illegitimate. Therefore, the next in line to the throne was his catholic brother, James. Inside parliament a faction known as the Tory Party openly acknowledged James as the future king. However, opposing them, on religious grounds, was the Whig Party; who were equally keen to obstruct James' accession. The resulting exclusion crisis bequeathed to the British Isles a permanent constitutional legacy in the form of multi-party politics.

Upon Charles' death, in 1685, his brother, in spite of the prevailing controversy, became James II of England. Many thought that James' destabilizing influence would end with the accession of one of his protestant daughters, Mary or Anne. But when James fathered a Catholic son, fears mounted. Capitalizing on those fears, his daughter, Mary, together with her protestant Dutch husband, William of Orange, invaded England – usurping James, who fled to France. This 'glorious revolution' (1688-89) created in Great Britain a truly parliamentary monarchy – one in which parliament became a permanent institution. All catholics were thereafter debarred from the succession, as the country became unwaveringly Anglican, i.e. Church of England.[47]

Many have interpreted this Anglo-Dutch merger, which witnessed the founding of the Bank of England in 1694 (and with it the establishment of a national debt due to government borrowing), as the origins of the British imperial war machine. A two-party system then dominated British political life for two hundred years. Such that by the time the Tory Party became the **Conservative Party,** in 1830, and the Whig Party had reinvented itself as the **Liberal Party,** in 1832, Great Britain had experienced astonishing changes to its entire infrastructure, economy and way of life. The driving force behind this wholesale change was the Industrial Revolution (circa 1760-1840), which was characterized by increased urbanization, mechanization and manufacturing.

However, such industrialization tended to create extremes of wealth and poverty within sprawling urban centres. With the group most affected by all this unplanned urban expansion being the working-class, whose life-expectancy fell to below half of that of the healthiest agricultural districts. Such marked inequalities led to rising concern amongst the socially aware – for example, the political philosopher Karl Marx (1818-83) viewed the state as an instrument of oppression and predicted class warfare. Marx collaborated on the **Communist Manifesto**, which envisaged the overthrow of the **capitalist system** (the system which lay behind the benefits and shortcomings of the Industrial Revolution). In place of capitalism he envisaged a state owned by the people themselves, with the wealth accruing to that state concentrating in the people's hands.

Marx's ideas, many of which were formulated whilst he was living in London, were mostly disowned by the British establishment, but were nonetheless seized-upon by Russian revolutionaries, such as Vladimir Ilyich Lenin (1870-1924). Lenin, as though taking his cue directly from Marx, urged the **proletariat** to seize power in Russia, following the successful Bolshevik Revolution, which took place in October 1917. After Lenin's death, all such power began to concentrate in the hands of Joseph Stalin (1879-1953), who swept away every concession to capitalism, and engaged in great purges, show trials and summary executions. Such was the punishing nature of unbridled Marxism.

The British saw this move towards communism in the east, be it Marxist-Leninism, Stalinism, Titoism, or any other variation on the same, as a premature abandonment of the benefits of capitalism. Any perceived advantages, associated with collective ownership, appeared to be overwhelmed by the inherent shortcomings of the one-party political system. At best, the working class, or proletariat, might become stakeholders in their own oppression. Moreover, the capital-owning middle-class, or **bourgeoisie**, became an "enemy of the people", and was therefore likely to be destroyed *en-masse*. This invariably led, in the communist Second World, to the kinds of abuses referred to in George Orwell's novel Nineteen Eighty Four.

Wisely, Great Britain avoided drifting to the political Left; but did concede that marked disparities of wealth were unacceptable. Consequently, the Conservative Party and Liberal Party were joined in 1906 by a third political force – the **Labour Party** (which was based on **socialist principles** and sought to represent the working class). As the Liberal Party's fortunes began to wane, following the start of World War I, politics became dominated by the other two parties, in an arrangement known as the Westminster Model. As for the **aristocracy**, that gave way to a burgeoning bourgeoisie, or middle class, due to the demands placed on British politics by socialists and reformers. Reforms which avoided the worst accumulations of wealth experienced in the preceding centuries, and which crucially extended the **franchise** or vote to women in 1918 (and to all of full-age in 1928).

Following the Second World War the 1948 Nationality Act reaffirmed, to all Commonwealth citizens and colonial subjects, their right of entry into the United Kingdom without restrictions.[48] The resulting influx caused some to question this policy, leading in turn to Enoch Powell's (1912-98) 'rivers of blood' speech – which predicted racial war. Racial war there wasn't, but there was certainly a great deal of cultural and ethnic diversification, as immigrants sought to enter Britain prior to fresh legislation – which, from 1962 onwards, began restricting both primary and secondary immigration. From the 1970s, however, the emphasis shifted very much towards European politics. So much so, that as immigration from former colonial possessions dried up, future immigration from Europe seemed increasingly likely.[49]

As well-balanced societies don't split hairs over minor differences pertaining to traditions, values and social norms – and as culture has served to create only superficial differences between the races – there's every reason to entertain such diversity. If anything, the escalation in unusual, strange, and often difficult to classify, physical and psychological phenomena in those decades, simply highlights the inherent similarities. By the late 1970s the world was entering a Second Cold war, and with it an increasing spiral of silence as regards telepathically-induced effects. Amid this climate of helplessness, conformity and acquiescence lay the routine manipulation of school-leavers – all of whom risked becoming yet more 'bricks in the wall' of strategic deception.

Within this environment of heavy-handed interference, military apprehension, floating exchange rates and monetary inflation there appeared a breakaway group of disgruntled Labour Party members, determined to offer the electorate governance from closer to the political Centre. This group briefly formed the Social Democratic Party (SDP), which allied itself to the Liberal Party, prior to the two parties merging in 1988, and adopting the name the **Lib Dems**. As for the Labour Party, that reinvented itself under the temporary banner **New Labour**; a left-of-centre party, cleansed of militant left-wingers. Just as the proletariat had appropriated somewhat bourgeois pretensions, so too had the Labour Party, which won the

158

1997 general election with a landslide victory, thanks to the popular leadership of Tony Blair (1953-).

As UK Prime Minister, Tony Blair fully supported US Republican President George W Bush's (1946-) controversial military action against Iraq in 2003. Anglo-American relations were particularly strong at the turn of the 21st century; cemented as they were by contentious military action overseas, a mutual determination to fight terrorism, and a sense that this **special relationship** was being decidedly pro-active, rather than simply laissez-faire. The darker side of this growing reliance on foreign military adventure and the attendant cult of militarism being shocking reports of abuse by US soldiers, and the casual abandonment of due process in respect of terrorist suspects. NEURON is unlikely to reward entirely laissez-faire behaviour; but neither will it tolerate abuses, perhaps even coming to the assistance of those who challenge the same.

As for Marx's concerns about mounting inequalities and the desperate need to avoid poverty amid advancing wealth: the answer appears to lie in capitalism's efficient creation of wealth – wealth which can be fairly redistributed.

- **America's mainstream revolution**

Following America's rediscovery in the 1490s colonization became unstoppable. By the time of the **American Revolution** (1765-89) there were 13 English-speaking colonies on the eastern seaboard of what is now the USA (see Fig 21). These colonists resented British control and fought hard to replace it with self-rule, the resultant **American Revolutionary War** leading to the desired self-governance. As for the representative body which had managed the war on the colonist's behalf – the so-called Second Continental Congress – that became the *de facto* national government of the US. At the country's inception, on the 4 July 1776, the government issued the **Declaration of Independence**, which stated: "we hold these truths to be self-evident, that all men are created equal, that they are endowed by their Creator with certain unalienable Rights, that among these are Life, Liberty and the pursuit of Happiness".

Soon afterwards the articles binding the thirteen states were replaced with the **US Constitution**, and the site of the eponymously-named capital was chosen (i.e. Washington D.C). Almost immediately a two-party political system arose, comprising the Democratic Republican Party and the Federalist Party. America's first President, George Washington (1732-99), a Federalist, favoured strong central government; unlike the Democratic

Republican Party, which was more concerned with state rights and limited government. In many respects Washington's instincts were correct, as the Union began to buckle under the weight of regional differences. Differences so great that the southern slave states would eventually test the might of central government through a policy of secession.

Figure 21: The settlement of America by white Europeans (1600-1861)

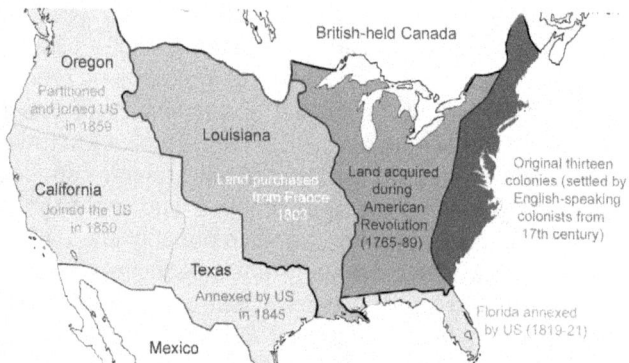

As the Federalist Party went into terminal decline, the Democratic Republican Party's position became ever stronger (changing its name, in 1828, to the **Democratic Party**). The issue of slavery and secession finally split the Democratic Party and a new party was formed, in 1854, called the **Republican Party**. The Republican Party was anti-slavery and opposed to the secession of the slave states. In 1860, Abraham Lincoln (1809-65), a staunch Republican, was elected President. Realizing that their place within the Federal Union was weak, the eleven southern slave states withdrew; forming the Confederate States of America, presided over by Jefferson Davis (1808-89). Unavoidably, these two sides became drawn into a protracted and bloody civil war as the Republicans fought uncompromisingly to preserve the Union, ostensibly under an anti-slavery banner.

The **American Civil War** (1861-65) was a pivotal moment in US social, political and economic history. Militarily, both sides possessed strengths and weaknesses. For example, the Union side in the north had a larger population, international recognition and a strong industrial base. Whereas the Confederate states, in the south, were able to put a greater proportion of their white population under arms due to unpaid slave labour. Fortunately for Lincoln, Ulysses S Grant (1822-85) rose to become overall commander of Union forces. His aggressive pursuit of Union victory and sheer tactical genius gave the north full control of the Mississippi River, effectively isolating the Confederate states to the west (see Fig. 22).[50] This, together with a successful naval blockade further weakened the south, forcing the confederacy to adopt bolder tactics.

Figure 22: America at the time of the Civil War (1861-65)

Audacious maneuvering on the part of Robert E Lee (1807-70), the commander of Confederate forces, led to the capture of Gettysburg, a small town in Pennsylvania. Three days of intense fighting followed as 150,000 men fought for control of this otherwise insignificant Union settlement. In an episode eerily prophetic due to its forlorn use of artillery, prior to a full-frontal assault, 'Pickett's Charge' was a bold gamble on Lee's part. Huge losses, however, led to a long slow rout of confederate forces. By 1864 the Union side was irrefutably winning the war, and this brought Lincoln resounding success in the Presidential elections held in that same year. Robert E Lee finally surrendered to Ulysses S Grant on April 9th 1865 (in a Virginian Court House). Six days later, Abraham Lincoln was dead, killed by an assassin's bullet.

Lincoln's successful emancipation of black slaves echoed Britain's abolition of slavery earlier in the century. The subsequent 15th Amendment to the US Constitution, introduced in 1870, stated that the right of a US citizen to vote should not be denied on account of race, colour or previous condition of servitude. However, black voters often found themselves cynically disqualified for reasons other than skin colour, which meant that America's Constitution appeared sound, but not so the actions of many of its citizens. In

161

fact, it would take an entire century and the charismatic involvement of Martin Luther King Jr (1929-68) before the **Voting Rights Act** (1965) was introduced, which did little more than echo the language of the 15th Amendment.

Women faced a similar uphill struggle for equality. In 1890 the two principal women's movements merged to form the National American Women's Suffrage Association. Later, in 1917, they were joined by the National Women's Party. Their combined efforts were eventually rewarded when the government introduced the 19th Amendment to the US Constitution, which stated that the right of a US citizen to vote should not be denied on account of sex. Although the 19th Amendment was law throughout the US in 1920, ratification by many states was unacceptably slow. Therefore, the phenomenon, which emerged between 1913 and 1947, deliberately positioned itself above the highly-misogynistic and overtly-racist national and subnational mainstream elements plainly visible within the country which had produced it. This made it nominally supranational from the start, almost by default.

Funding this neuro-cognitive power also became easier as the civil war years had forced a wholesale rationalization of the US banking system, with the National Banking Act, of 1863, creating 1500 national banks – which, together, provided for a uniform currency. However, persistent problems within the banking sector nationally (such as 'liquidity crises') led in 1914 to the establishment of the present Federal Reserve System. Between them, the Treasury Department and the Federal Reserve System undertake the duties normally associated with a central bank.

The **progressive era** (circa 1890-1930), as it became known, spanned those years when America underwent a significant metamorphosis. Prior to this era most of the emphasis had been on: 1) reconstruction following the civil war; 2) introducing a uniform currency; 3) black emancipation; and 4) the creation of large planned urban centres. Building on this, the progressive era witnessed a shift towards immigration, industrial production, women's suffrage, electrification, radio-communication, movie-making and the promotion of the now ubiquitous car-culture. However, the infrastructure of modernity proved easier to fashion than the psychology of those occupying the same, with disparate social elements clashing uncomfortably within this new cosmopolitan environment. Thus, real progress would take far more than a mere 40 years to materialize, as the federal authorities began managing the slow decline of organized crime and the **Ku Klux Klan**.

Figure 23: Republican and Democratic Party
Presidents (1861-1961)

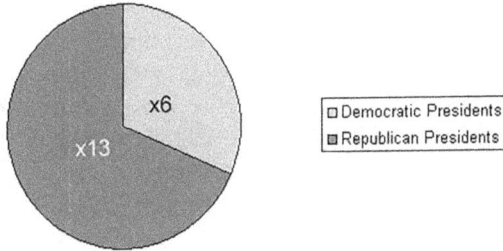

The dominant political party at this time was the Republican Party (see Fig. 23), which helped to forge a cohesive United States of America by strengthening the hand of federal government; not least of which through its proposed banking and currency reforms. In fact, the 16th Amendment, ratified in 1913, provided for the collecting of taxes "from whatever source", at a time when science and technology would prove decisive, both militarily, politically and commercially. Characteristically, the United States has attempted throughout its rich history to codify its good-intentions, through amendments to its original constitution. Contemporaneous with this has been the presence of unruly elements within its own population – which only modern constitutional apparatus has the potential to effectively contain.

Today, the Republican Party is associated with a Right-of-centre agenda, emphasizing pro-American foreign policy, rule of law and military superiority; while the Democratic Party is perceived as Left-of-centre, with the emphasis on social reform, civil liberties and the pursuit of sound energy policies. For example, John F Kennedy, a Democrat, pursued liberal reforms and avoided escalating the war in Vietnam; whereas Ronald Reagan, a Republican, proved vociferously anti-Communist and introduced the Strategic Defence Initiative. It is, in fact, demonstrative of just how far Britain's Labour Party has moved to the political Right, becoming only slightly Left-of-Centre, that it was able to form such an effective alliance with the Republican Party, following the events of 9/11.

Succeeding the Republicans, in 2009, was the Democratic Party, and America's first ever African-American president, Barack Obama (1961-). Which just goes to show that America's mainstream revolution is alive, ongoing and unstoppable.

• Elective dictatorships

The phenomenon responsible for telepathically-induced effects was born out of the US Constitution, sufficient to bring us to where we are today. But that sets it apart from subnational and even national politics, which were ridden with inertia as regards much of what the constitution proposed. The true picture is further complicated by, amongst other things, male-domination of its internal structure, its determination to enervate the extremes of Left and Right, and the absence of a clear remit following the collapse of communism in the east. This one hundred-year revolutionary war (1913-2013) has been waged by sea, land and air; and with all the strength that an invisible neuro-cognitive power could muster. Of course, an observer might fairly ask: where has that hundred years taken us? This is an important question, lest we introduce the politics of the disastrous, in place of the politics of the unpalatable.

As we've seen the political systems in Britain and America both possess an executive who plans policy, a legislature which enacts laws, and a judiciary to decide on cases arising as a result of those laws being allegedly broken. In the US the executive is the **President**, the legislature is **Congress** (comprising the Senate and the House of Representatives), and the judiciary is headed by the **US Supreme Court**. Separation is so complete that the president cannot serve in Congress; nor can a serving Congressmen become a Supreme Court Judge (their respective roles being defined by the US Constitution). In Great Britain the executive is the **Prime Minister** (and their Cabinet), the legislature is **Parliament** (comprising both the House of Commons and the House of Lords), and the judiciary is headed by a **UK Supreme Court**. The machinery of decision making in the UK is further complicated – even today, post-Brexit – by the influence of Brussels, by which we mean the **European Union** (EU).

In Britain the constitution is **uncodified**, in-so-far as there is no single written constitution. At best, what exists is only partly written, e.g. statute law, common law, conventions, etc. This makes the legislature the unchallengeable authority in constitutional matters – in other words, parliament becomes the ultimate arbiter.[51] Even when a mainstream constitution is **codified**, as in America, it's not wholly immutable, and can be subject to an unspecified number of amendments by Congress. Thus, significant control of the legislature makes the constitution vulnerable to the democratic process, such is the nature of democracy down at the national and subnational levels. The ruling party could, on occasions, be constrained only by the need to win future elections – hence the term, **elective dictatorship.**

Far from amplifying this potential for elective dictatorships a supranational presence, possessing a neuro-cognitive capability, would act as a counterpoint to the same. We need to ask ourselves, however, at precisely what point would NEURON begin to compensate for errant mainstream behaviour? We've already seen how Adolf Hitler's rise to power began with a variation on elective dictatorship; in-so-far as the Nazi Party became the largest in the Reichstag, and was able to pressure the same into passing the Enabling Act (1933). An act which fundamentally altered the constitution of Germany; and in a manner which granted Hitler absolute power, concentrating power not simply in the hands of the party, but in the hands of a given individual.

One might be facilitating that slide into totalitarianism by merging the subnational, national and supranational elements too much. Which both explains and justifies NEURON's commanding position. In reality, political pathology can be challenged at either the subnational, national or supranational level. It just so happens that supranational answers carry with them the potential for death and trauma. The onus is therefore on mainstream decision makers to manage political pathology at the subnational and national levels, or face the prospect of a supranational solution presenting itself.

- **Postwar welfare provision**

Great Britain's Emergency Powers Bill (May 1940) empowered the government to direct anyone to do anything in the national interest and to assign any property to any national end that it chose. Therefore, by the time Sir William Beverage published his famous report into postwar welfare provision, in November 1942, his recommendations must have seemed relatively decentralizing. In Britain, following an Allied victory, Beverage expected to see the following: 1) National Health Service; 2) state pension; 3) family allowances; and 4) near full employment. Beverage's proposals served to create a light at the end of the tunnel, as the war began to turn in the Allies favour. In that sense they were a well-balanced incentive, which seemed to be saying "keep going".

Those of us living in the UK are still benefiting from these welfare provisions. Unfortunately, the ill-balanced application of telepathically-induced effects has the capacity to force people out of work, drive up the demand for welfare payments, make providing for one's old age more difficult, and all-the-while whilst breaking-up families. This is why understanding this dynamic is so important to the western economies, and why complacency on these issues is so reckless. What's more, if the state were to withdraw welfare provision, whilst turning a bind-eye to unruly interference, it would be asking people to compensate for the absence of financial support, whilst hampered by externally-Imposed difficulties.

In the USA welfare provision became a prominent issue during the Great Depression (1929-1941), when private charities, state departments and local governments became overwhelmed by the sheer number of people facing acute poverty. Today, the Federal Government does much to support individual states in meeting those needs; but given that the cost of welfare could well exceed $1,000 billion per annum in the near future, the onus is now on those applying radiofrequency effects not to unduly aggravate the impact of poverty. This is another reason why getting the balance right is so important, because excessive interference may serve to swell the ranks of those on the margins of politics.

- ### Devolution, separatism and secession

Some of the world's most intractable problems relate to secession and separatism. And, indeed, the United Kingdom has had its share of such issues, Westminster having ceded power to devolved parliaments and regional assemblies in Northern Ireland (1998), Scotland (1999), Wales (1999), and London (2000). In the case of Ireland, the entire land mass was bound to Great Britain by the Act of Union, in 1801. Thereafter, tensions simmered until the outbreak of the **Anglo-Irish War** (1919-1921), which led to partition – hence the Republic of Ireland in the south and Northern Ireland in the north. The short-lived civil war, which followed, did little more than furnish us with **Sinn Fein** and the **IRA**, organisations which reasserted themselves after 1969.

Then, in 1984, one and a half decades after the troubles in Northern Ireland had resurfaced, the 'Brighton Bomb' very nearly killed the British Prime Minister, Margaret Thatcher, during an attack on her cabinet. Further outrages prompted her successor, John Major (1943-), to exclaim: "we are determined to do all we can to bring peace...further killings make that search for peace all the more urgent". The subsequent peace process, which was initiated by John Major and finalized under his successor, Tony Blair, survived numerous setbacks. But eventually, the Good Friday Agreement was signed, and a Northern Ireland Assembly was provided for at Stormont (1998). This being a prime example of elective power being exercised in a particularly constructive manner.

Unlike Ireland, there was no recent history of bloodshed in respect of Scottish autonomy. James II's grandson, popularly known as Bonnie Prince Charlie (1720-88), had the support of many in Scotland following the dissolution of the Scottish parliament by the British. But the wholesale destruction of the Scottish Jacobite cause, as it was known, led to Scotland becoming, if anything, an active participant in British imperial ambitions from the late 18th century onwards. Thus, the British Isles were spared Scottish sectarianism, and, instead, were treated to the very best of what Scotland could produce, be it economist Adam Smith (1723-90), philosopher David Hume (1711-76) or novelist Sir Walter Scott (1771-1832). A lesson to all Eurosceptics that participation, not confrontation, can actively benefit both parties. Scotland did eventually have its parliament returned, in 1999; which is contemporaneous with the establishment of assemblies in Wales and London.

Significantly devolution returns decision making power to a region, but not sovereignty; whereas federalism gifts sovereignty upwards to a transnational body. This process of **devolution** coincided with an energetic debate regarding the importance of **subsidiarity**, i.e. the belief that political and municipal functions should be carried out at the lowest practical level. Subsidiarity implies that if a function cannot be adequately performed at the subnational or national level then one must invent a supranational solution. Such solutions would invariably transcend the states which produced them.

Sovereignty dictates where mainstream power ultimately resides, with the sovereign state being perceived as a person in the eyes of international law, replete with the equivalent of a self-schema, i.e. the total set of memories, ideas and intentions that it holds about itself. This obviously provides for attributional questions arising in respect of states – and, indeed, fundamental errors.

• The rise of Western Europe

Ratifying the Final Settlement in Respect of Germany was one of the last official acts by Mikhail Sergeyevich Gorbachev, prior to the formal dissolution of the USSR. So ended every semblance of the Cold War, that state of political and military tension which had existed for more than 40 years between the capitalist First World and communist Second World. To understand just how much Western Europe had evolved in that time, we must rewind the clock to 1949. That was the year in which the North Atlantic Treaty Organization (NATO) was formed: its signatories being the USA, Canada, UK, France, Denmark, Italy, Norway, Portugal, Iceland, Belgium, Luxembourg and the Netherlands. In principle, an armed attack on any of these countries would represent an attack on them all. Therefore, NATO provided significant nuclear protection to the whole of Western Europe.

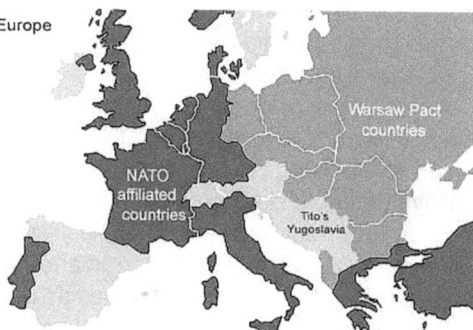

Figure 24: Western Europe in 1955

With military security provided for, the West began the long slow process of rebuilding Western Europe (see Fig. 24). Today, we take it for granted that Europe has coalesced into a community of nations; but in reality, NATO, on its own, would have been sufficient to deter Soviet aggression and foster mutual relations. Instead, a monumental decision was made to embark on an altogether more prophetic, poignant and significant alternative – the crafting of a single European market of goods, services and people; arrived at through the piecemeal relaxation of borders, and the surrendering of so much sovereignty. Such a community began to emerge in the 1950s, comprising France, Italy, the Netherlands, Belgium, West Germany and Luxembourg. Between them, they set-up what was colloquially known as the **Common Market**, via the Treaty of Rome of 1957.

This European Economic Community (EEC), to give it its more imposing title, was later joined, in 1973, by Denmark, Ireland and the UK. This was the year when floating exchange rates and the first of two oil crises began to adversely affect all the major European economies. Several years later, in 1979, after being ravaged by monetary inflation, rising import costs, and increased currency speculation, a **European Monetary System** (EMS) was established. Next came the **Single European Act** (SEA), which provided for the free circulation of goods, services and people within the EEC. German reunification, and the disintegration of the USSR, then led to a wholesale review of the **European Community** (EC).

Figure 25: Western Europe in 2004

NOTE: NATO affiliated EU and Non-EU states dominate Western Europe - more so, since the accession of Romania and Bulgaria in 2007. This has served to increase tension in the Ukraine (and arguably Belarus) leading in 2014 to a state of instability and disorder in the region.

What followed was the Maastricht Treaty (1991), which effectively swept away the EC, which was little more than a prototype, and replaced it with the European Union (EU). The establishment of the EU led to a whole new monetary system, called **European Monetary Union** (EMU); which began issuing *euro* notes and coins in 2002. Shortly afterwards, the EU ratified the **Treaty of Athens** (2004), which increased the number of member states from 15 to 25, through the incorporation of many former Warsaw Pact nations, i.e. countries which had once been allied to the USSR (see Fig. 25). Consequently, there has been both a deepening and a widening of this particular union – a union which looks well-placed to deepen and widen some more (as evidenced by the accession of Bulgaria and Romania in 2007).

Post-Brexit, Great Britain continues to cling tenaciously to the pound (not surprising given that it is a truly international currency), making its subsequent abandonment all the more significant, if that ever happens. This book perceives the free world as comprising those countries which are affiliated to NATO (with NEURON spanning the entire geopolitic). To be bound to NATO, and subordinate to NEURON, is of prime importance if a country is to benefit from extrasensory effects. The importance of the European Union lies in the fact that it creates a rigid power bloc, which can't help but attract the interest of neighbouring countries, thereby expanding the limits of the free world. For these reasons, the European Union is of immense importance – irrespective of whether the United Kingdom abandons itself to the same – and its discontinuous development ought to be encouraged.

Ch14: Global politics

- ### The Far East

The **Second Sino-Japanese War** (1937-1945) witnessed the Japanese invasion of China, and with it a whole litany of atrocities (e.g. rape of Nanking). America responded punitively to Japanese expansionism, and the associated atrocities, by imposing an embargo on the export of aviation fuel and metals to Japan. This well-balanced response to Japanese hegemony led, in turn, to the Japanese attack on **Pearl Harbour** (7 Dec 1941). Over the following days the Allies all formerly declared war on Japan, including the principal victim of Japanese aggression, China. China's nationalist leader, Chiang Kai-shek (1887-1975), became an effective military ally of the West, receiving much help and support from the United States. When the Japanese finally capitulated, following the dropping of the atomic bomb, the Chinese then descended into an intense period of civil war (1946-1949). These were the years in which the Americans and Soviets both sought to make the peace in the Far East, through political and military maneuvering.

Thus, the Chinese Civil War witnessed nationalists, under Chiang Kai-shek, fight communists, under Mao Tse-tung, for control of mainland China. In the event, the communists won, proclaiming mainland China to be the **People's Republic of China** (PRC). As for the nationalists, they retreated to the island of Taiwan, where Chiang Kai-shek established what he subsequently claimed was the legitimate government of China – making Taiwan the home of the Chinese government in exile, i.e. the **Republic of China** (ROC). Some would attribute the communist's success to Chinese proletarian leanings, allied to naiveté at the vagaries of the one-party system. Others would ruefully point out that the nationalist's failed to appear sufficiently pro-American, and therefore wholly misread the new world order.

The Soviet Union immediately recognized Mao Tse-tung's People's Republic of China, making China and Russia the backbone of an emergent communist Second World. The pervasive influence of the two prevailing **superpowers** (namely, America and the Soviet Union) meant that local schisms would often appear on the ground, due to the presence of pro-American and pro-Soviet factions. It was within this climate that both Korea and Vietnam became partitioned. Thus setting the scene for the ensuing **Korean War** (1950-53) and **Vietnam War** (1964-75). American capitulation in Vietnam in the mid-1970s, amid burgeoning oil prices and economic misery, all served to underscore the cost of war – politically, financially, psychologically, emotionally, and in terms of the sheer numbers of lives lost.

The nature of Vietnam today suggests that the real winners of the Vietnam War weren't the **Viet Cong** (i.e. communist Vietnamese), but rather the **Viet Minh**, who were principally nationalists hell-bent on autonomy and self-governance. For these nationalists any association with the communist Second world was a mere convenience. Additionally, the Vietnamese, or Viet Minh, went on to displace the Marxist-Leninist **Khmer Rouge** in Cambodia, suggesting major ideological differences. Therefore, North Vietnamese leader Ho Chi-Minh (1892-1969) achieved his principal aim; which wasn't communism, but rather self-governance – to him the Vietnam War was a war of independence. But maybe America's own inner-voice was becoming conscious of that fact, making ultimate capitulation that much easier to stomach.

Figure 26: The Far East

While the Vietnam War entered into American folklore as a defeat, the Korean War which preceded it ended in an unmitigated stalemate. The Korean Peninsula today is still divided between a decidedly pro-American South Korea and a pro-communist North Korea (see Fig. 26). However, with the collapse of the USSR, and the continued liberalization of mainland China, North Korea risks becoming a politically-marooned rogue state. As for Tibet, the province invaded by mainland China in 1950, its future remains uncertain because it's unknown for the Second World to hand sovereignty back to a region without a wholesale collapse of the communist system. We can see evidence of this in Mongolia, which came under direct Soviet control in the 1920s, but which made the transition to a multi-party democracy in 1992 with comparative ease, following the dissolution of the USSR. Communist China itself remains controversial, for, although the Soviet Union immediately recognized the People's Republic of China, the United Nations took far more convincing – only belatedly allowing mainland China to assume China's seat at the UN, in 1971.

As for Japan, following its unconditional surrender to American forces at the end of the Second World War it eventually had its sovereignty restored, in 1952 – improbable, if it had surrendered itself to the Soviets or Chinese. Thereafter, Japan re-invented itself as a formidable economic power, deriving much of its revenue from the engineering, automobile and electronics industries. Due to its voting system it has tended to produce elective dictatorships. Thus, the Liberal Democratic Party of Japan has all but dominated postwar politics – not least of which due to the overall strength of the Japanese economy. However, with the economic tide now turning, and the world recession beginning to bite, a 'two-party system' may be in the offing. So, whilst the communist world has suffered disintegration and collapse; and whilst the neutral or non-aligned nations have endured so much stagnation; the pro-western democracies have consolidated their gains (and even helped allies into existence).

The **Four Asian Tigers** (Hong Kong, Taiwan, Singapore and South Korea) are, or have been, closely associated with the West. Hong Kong, for example, was held by the British until the expiration of its lease in 1997, whereupon it fell under the control of mainland China; Taiwan is a multi-party democracy with universal suffrage, which counts the US as a principal ally; Singapore is a city-state and Commonwealth member, with a parliamentary system based on the Westminster Model; and South Korea has close ties to the European Union, the country's single largest investor. These are all centres noted for their rapid industrialization and exceptional growth-rates. Singapore, for example, holds the distinction of being the fastest growing economy in the world.

Other countries in the far east, such as Burma (otherwise Myanmar), Malaysia, Indonesia and the Philippines – all with longstanding histories of colonial rule, often involving multiple imperial powers – actively sought (and ultimately achieved) independence following the last war. These former imperial possessions have tended towards nationalism, and a strong sense of regional identification. Consequently, in 1967, Malaysia, Indonesia, Singapore, Thailand and the Philippines jointly formed the Association of Southeast Asian Nations (ASEAN). Thus replacing the hard heteronomous power of colonial rule with the soft power of direct foreign investment, much of it coming from the European Union, Japan and the US. Since then ASEAN has been joined by Burma, Cambodia, Laos and Vietnam.

The decline of British colonial rule in the region coincides with the establishment of one of the United Nations principle bodies – namely, the Trusteeship Council (which managed the granting of independence to former colonies and imperial possessions following the end of the war). It was within this more progressive postwar climate that the British Empire faded gradually into an association of equal partners, termed the **Commonwealth of Nations**. All Commonwealth nations are bound by the **Singapore and Harare Declarations**, which commit the signatories to certain shared ideals, such as democracy, human rights, good governance and rule of law. Presently comprising 54 countries, the Commonwealth includes the UK, Australia, Canada, New Zealand, Singapore and Malaysia.

Thus, the countries of the free world – and the UK, USA and EU in particular – have all fostered important strategic links with the Far East. So much so, that if this ultimately serves to fatally wound communism in the region, leaving Tibet unmolested, so be it.

• Global alliances, leagues and associations

Existing intergovernmental partnerships would be observed by NEURON in the interests of no single nation, just one single whole. Intergovernmentalists favour active cooperation between governments – in ways which are mutually advantageous, but without the surrendering of any sovereignty. Examples of the same include the Commonwealth, North Atlantic Treaty Organization (NATO), United Nations (UN), and the Intergovernmental Panel on Climate Change (IPCC). Overlap is rife, with sovereign states often bound together in multiple ways through a plurality of consortiums, confederations and bodies – each committed to a variety of social, political and economic goals. Extrasensory effects provide for translation and interpretation, and could even profitably enrich the information received in the course of such exchanges.

The **Intergovernmental Panel on Climate Change**, for example, was set-up in 1988 by the World Meteorological Organization and the UN to scrutinize and assess the scientific evidence regarding climate change. Today, it comprises hundreds of scientists from many different countries. Its conclusion is that greenhouse gas emissions from human activity (principally CO_2 emissions from industry) have resulted in an anticipated rise in global temperature – perhaps as much as several degrees Celsius by the end of the century. Whatever the true figure, anticipated warming is likely to raise sea-levels, acidify water as CO_2 dissolves, increase the frequency of extreme weather events (due to energy trapped in both the atmosphere and oceans), and cause the world's climatic zones to shift. NEURON would therefore assist those responding in a conscientious manner, and facilitate effective communication between the same.

Issues such as climate change, nuclear weapons proliferation and energy crisis (all being significant threats to global peace and security) can be broached more effectively with the advantage of extrasensory effects. Every climate – be it economic, strategic, political, military or environmental – has the capacity to undermine global peace and security, and so these alliances, leagues and associations help nations to avoid, or at least manage, the impact of such stresses through the dialogue they create (a dialogue which could be facilitated through telepathically-induced effects). Ultimately, stress is often defined in strictly economic terms, and so many of these alliances, leagues and associations constitute an attempt at offsetting the blunt trauma of financial disruption.

Figure 27: Asia-Europe Meeting (ASEM)

We've already seen how **ASEAN**, i.e. the Association of Southeast Asian Nations, comprises several of the smaller nations of Indo-China, specifically those sandwiched between China to the north and Australia to the South. Eclipsing ASEAN, in terms of geographical spread (see Fig. 27), is the much more expansive

ASEM (Asia-Europe Meeting). ASEM focuses on issues such as politics, security, education, economics and culture; and includes the entire European Union, Russian Federation, People's Republic of China, ASEAN and Australia. The land area covered by ASEM is therefore colossal, and crosses many different divides, be they religious, ethnic or ideological. Conspicuous by its absence from ASEM is North Korea, whose principal ally, China, might yet liberalize as a result of its relationship with Europe, Australia, Singapore, etc.

Prima facie the European Union and the United States of America appear to compete for international primacy – but such primacy is, in fact, a role they could comfortably share. This is because through their membership of NATO they remain formidable strategic allies. The principal difference between the European Union and the United States lies in how they relate to one another economically. This book proposes that Europe and America accommodate whichever economic arrangement serves the free world. With these two regions then competing for the advantages conferred on them from above – the most ideal possessing the greatest extrasensory advantages. In this context ideal equates with legitimacy, balanced-mindedness and democratization.

One can see evidence of this apparent competition in the proposed Trans-Pacific Partnership, or **TPP**. The TPP is a free-trade area presently comprising Brunei, Chile, New Zealand and Singapore; but is currently negotiating a huge expansion – one which would merge the whole of North America, Australia, Peru, and the TPP's existing members into a single market. The TPP therefore looks set to add to the competition which currently exists, possibly strengthening the hand of NATO in the process, through the demands it makes on human ingenuity, application and inventiveness. As for the UK, it could afford to abandon itself to a European Union competing to be the most legitimate power on earth. And, with the power of the free world concentrated in at least two geographical locations, some protection is afforded against the destructive effects of natural catastrophe, stock exchange crash and political pathology.

One of the principal reasons for Scotland, Northern Ireland and Wales remaining in Europe, apart from acting as a bridge between the USA and EU, is about access to global markets. The African, Caribbean and Pacific Group of States, or **ACP**, exists to promote sustainable development and poverty reduction in the Third World (see Fig. 28). Recent agreements between the EU and ACP have stressed the importance of mutual obligations in respect of human rights, democracy and the rule of law. Violating any of these conditions may result in the suspension of EU development aid. Thus, the profits accruing from investments in the Far East could, in theory, find themselves ploughed into enterprises in Africa, the Pacific islands and Caribbean – but only if those obligations are being met.

Figure 28: African, Caribbean and Pacific Group of States (ACP)

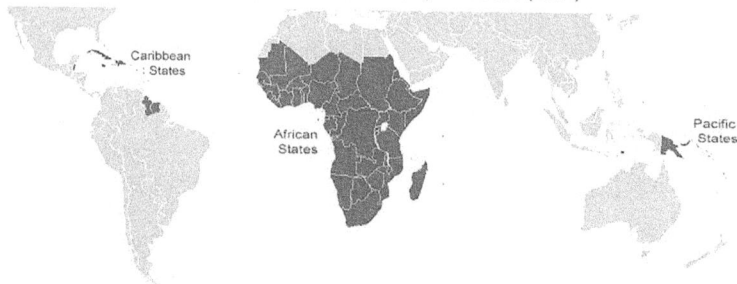

Singapore is a city-state possessing a parliamentary system based upon the Westminster Model, and therefore in many ways is deserving of active support and financial success. However, even when capital and influence concentrate in just the right hands, that still leaves legitimate powers at the mercy of oil shocks and energy crises. The oil consortiums which exist today do more than protect their own interests, they guarantee supplies to an energy demanding free world. But a free world which is nonetheless conscious of the need to maximize the rewards associated with modern living, whilst reducing as far as possible the punishing effects of modernity.

Supplying the energy needs of the technically-advanced free world – the geopolitic most committed to reducing the overall impact of population pressure, energy crisis, climate change, unacceptable poverty, nuclear arms proliferation and racial discrimination makes sense. Especially if, in the course of meeting those needs, the world becomes a safer, more secure and prosperous environment. Additionally, increasing the First World's dependence on renewable energy sources strengthens its hand when bartering with the major energy consortiums – the most conspicuous of which is **OPEC** (Organization of Petroleum Exporting Countries). OPEC presently comprises thirteen member states (five Middle Eastern, seven African, and one South American). Relations between the Middle

East and South America are further strengthened by the Summit of South American-Arab Countries, or **ASPA** (see Fig. 29).

Figure 29: Summit of South American-Arab Countries (ASPA)

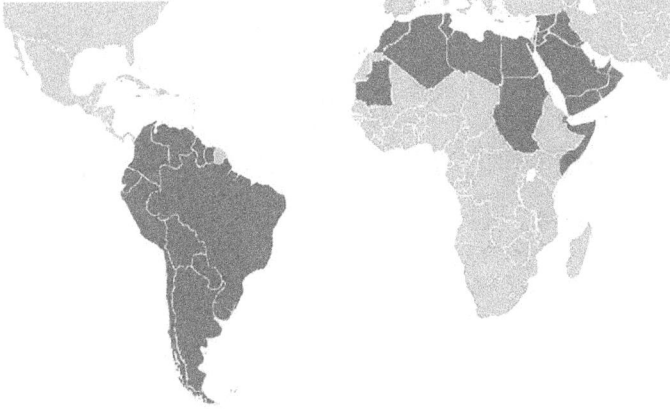

The United Kingdom is well-placed to support an overarching neuro-cognitive power professing a global reach, and provide feedback to the same, due to its complex rapport with other countries, balanced-politics, empathy with autonomous aspirations, opposition to the extremes of Left and Right, and longstanding commitment to the process of economic and social development worldwide. Such attributes more than justify its permanent seat on the United Nations Security Council. The five permanent Security Council members, or P5 (namely, USA, UK, France, Russia and China), were the immediate victors, or their chief allies, following World War 2. Of these, Russia and China have since experienced significant upheavals in their domestic politics, making them very different countries to the ones who won the war.

The real winners of World War Two are those who presently aspire to govern from as close to the political Centre as practicable, and who have made real telepathically-induced effects. That scientific expertise was initially drawn from Europe, America and the UK – with these regions making possible the next major phase of human progress, based on a model of their own choosing. If Russia joins Europe (and thus NATO) no less than four-fifths of the UN's permanent Security Council members will be signatories to that proffered manifesto. A manifesto which provides for democracy, free-will and progress worldwide. This raises the question as to what role the United Nations will serve in the future. The answer, of course, is that the UN is the final political instrument standing between mainstream factions and difficult and contentious supranational solutions.

• **The United Nations (UN)**

The United Nations (established, October 1945) comprises six principal instruments: 1. **General Assembly**; 2. **Security Council**; 3. **Secretariat**; 4. **International Court of Justice**; 5. **Economic and Social Council**; and, 6. **Trusteeship Council**. Arguably, the most important of these is the Security Council, which is dominated as we've seen by its five permanent members, all of whom are nuclear powers (and the only nations on earth permitted to possess nuclear weapons under the Nuclear Nonproliferation Treaty, which came into force in 1970). These P5 states (currently comprising the USA, UK, France, Russian Federation and Peoples Republic of China) all hold the balance of power in the assembly through their right of veto; which enables any one of them to block a draft UN resolution. And, although this can result in a continuation of localized conflicts, it does, nonetheless, help to preserve peace and security in the world.

It was anticipated that this newly established body would be more effective at maintaining world peace than its ill-fated predecessor, the League of Nations – and, to date, that appears to be the case. The General Assembly (presently comprising 193 member states) regularly deliberates on international affairs. In 2012, a proposed state for displaced Palestinians was officially recognized by two-thirds of the General Assembly – those voting in favour of the same being principally those countries from the Second and Third Worlds (and Greater Middle East). Those nations which form the strategic backbone of the free world either opposed the same or abstained. In reality, the capacity of displaced Palestinians to engineer a city-state, comparable to Singapore, rests on it being entirely workable, and not simply reliant on others for its survival.

The Security Council, which has deliberated on the Arab-Israeli conflict ever since its inception, has the maintenance of international peace as its priority. The Secretary General, who is amply supported by the Secretariat, serves a largely diplomatic function (acting as ambassador, mediator, negotiator and envoy). As for the International Court of Justice, that has a jurisdiction extending to all the member states, but in practice is heavily reliant on their individual consent. And, the Economic and Social Council examines the social impact of global economic forces, and reports back on the same. Whilst the now largely defunct Trusteeship

Council has overseen the granting of independence to more than eighty former colonies and imperial possessions. The Allies had therefore successfully established an organization committed to dismantling both the British and French empires. By way of recompense, these two former colonial competitors, Great Britain and France, dutifully took their place amongst the 'big five' on the Security Council – furbishing the newly emergent Europe with an important power of veto and overt nuclear protection. Meanwhile, the USA pursued its own interests through the power of its own veto, sheer military muscle, overseas diplomacy and, more controversially, covert activities. Following the dissolution of the USSR the seat occupied by the Soviets was simply appropriated by the Russian Federation, just as the Peoples Republic of China had assumed China's seat twenty years earlier.

When the phenomenon responsible for neuro-cognitive effects first emerged in the US it positioned itself above unruly subnational and national elements; such that by the time of 9/11 it was wholly supranational in nature. It now looks set to subordinate the whole of the United Nations, but not necessarily in the interests of the US. The sacrifices of two world wars have therefore resulted in a significant global presence – one which transcends the whole of mainstream politics. Strangely egalitarian, it has self-evidently created an assembly of equals, every one of whom has unalienable rights as regards telepathically-induced effects. In theory, all who subordinate themselves to the same stand an equal chance of being rewarded with life, liberty and the pursuit of happiness.

- ## The Nuclear non-Proliferation Treaty (NPT)

The Treaty on the Non-proliferation of Nuclear Weapons (more easily enunciated as the **Nuclear non-Proliferation Treaty**, or NPT) came into force in 1970, and is a landmark piece of legislation which seeks to inhibit the spread of nuclear weapons across the globe. Only those countries who tested nuclear weapons before 1968 are officially recognized as nuclear weapons states (i.e. the P5 countries). All other countries are expected to renounce nuclear weapons technology altogether. The treaty does, however, provide for the peaceful use of nuclear technology; specifically in the area of energy production. In spite of the NPT, several other countries appear to have appropriated these weapons of mass destruction, presumably through the five permanent Security Council members.

In addition to the five recognized nuclear weapons states, other countries who are believed to possess a nuclear weapons capability

179

include: India, Pakistan, Israel and North Korea. India and Pakistan are both members of the Commonwealth and allied to the West (but remain mutual enemies). Israel, on the other hand, is surrounded by a host of anti-Western Arab factions, whilst being strongly allied to America. North Korea, for its part, causes the West the greatest concern of all, due to its affiliation with Communist China, which has, according to some estimates, the world's largest conventional military force. Additionally, Iran may have acquired the expertise needed to produce an atomic bomb through its nuclear program, the one it appropriated at the time of the Islamic revolution in 1979.

A neuro-cognitive capability is more worrying than a nuclear capability, and a limited exchange of battlefield nuclear warheads is less significant than a wholesale exchange of radiological interference. The United Nations, for its part, would continue to be a mainstream assembly committed to resolving such issues by orthodox means. Whether its Secretary General would be afforded special extrasensory powers remains an open question. In all probability such a capability would be a reward bestowed on Secretary Generals who excel, not a way of making poor ones look good. As history has shown, the UN's future reliance on such effects is greatly reduced due to power and influence concentrating in the right hands.

The enlightened application of radiofrequency effects wouldn't have arisen in the communist Second World, the developing Third World, or Greater Middle East. In these regions, the unruly use of power has all too often frustrated progress on human rights, sexual equality and personal freedom. So, in spite of the best efforts of the **International Atomic Energy Agency** (IAEA) the potential for localized nuclear conflict in these regions remains high; with uranium deposits being plentiful enough to furnish mankind with its most punishing mainstream capability well into the next century. As always, it's the "splinters of the splinters" procuring nuclear weapons materials in secret which concerns us most – but of course it wouldn't be entirely secret, and a Secretary General of note might wish to distance the chief protagonists (as one might have felt compelled to do with the co-conspirators Austria-Hungary and Germany in the run-up to the First World War).

- ### Globalization, peak theory and the world economy

Globalization is the process whereby goods, services, capital and people form part of a much wider integrated global economy. By its very nature, globalization means more than simply an increase in the number of international financial transactions, it implies something deeper and more integrated. This process of economic integration, which has its roots in the proliferation of free-trade areas amongst the various alliances, leagues and associations mentioned above, has accelerated due to a vast increase in the use of electronic money, satellite communication, 24 hour trading, and the internet. Fortunately for NATO, the US dollar, pounds sterling and the euro are truly international currencies; whereas most other currencies are purely local tender. Prima facie then, these powerful global economies, centred on Europe and America, have the potential to service the worldwide economy through their joint actions and behaviours.

The United Kingdom, European Union and America are all members of the **International Energy Agency** (IEA), an organization which arose in the wake of the 1973 oil crisis. This agency exists to anticipate, and therefore avoid, the problems arising when resources such as petroleum, natural gas, coal and uranium are physically disrupted; but its entire remit spans the whole of energy policy, energy security, economic development and environmental protection. Peak petroleum, peak gas, peak coal and peak uranium (by which we mean the year in which these naturally-occurring resources are expected to peak, in terms of production) is, according to most modern estimates, sometime in the 21st century. Thereafter, the prognosis is one of general depletion, wholesale decline, and increasing scarcity. Hence the importance of the IEA.

Thus, globalization, nuclear proliferation and the growth of NATO are all contemporaneous with this approaching peak in overall production. **Peak theory**, which was first proposed by American geophysicist M King Hubbert (1903-89), implies that after peak production is achieved one is then faced with a terminal decline in extraction, refinement and supply. The term **energy security** therefore pertains to the availability of natural resources and primarily the safeguarding of cheap and reliable forms of energy, so vital to modern economies. As the present pre-peak world passes into history and we are faced with a bleak unending post-peak future, we may find that legitimate powers attract less criticism, especially when drawing heavily on natural resources. They will, of course, attract even less criticism if they lead the world in energy conservation, environmental protection and sustainable development (drawing heavily on the power of the sun, wind, waves, tides and geothermal activity in the process).

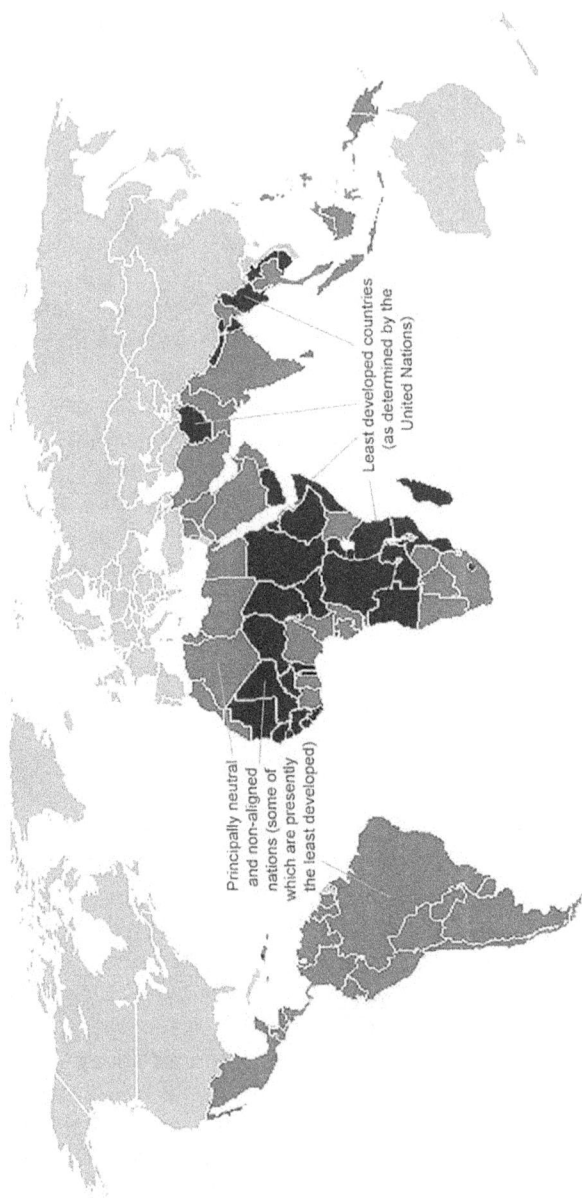

Figure 30: The neutral and non-aligned nations (and today's least developed countries)

Least developed countries (as determined by the United Nations)

Principally neutral and non-aligned nations (some of which are presently the least developed)

Peak theory pertains strictly to energy and resource production, but nonetheless has its corollary in 'peak population'; which will ultimately determine when shortage becomes critical. Economist Thomas Robert Malthus (1766-1834) argued in an essay on populations, that they naturally increase faster than their means of subsistence. One way to postpone that relative peak in energy production is to constrain demand. In other words, have fewer people. But that's only possible if we conceive of balanced-minded ways of avoiding over-population. After all, fewer people means fewer people to experience the pain and disruption of global energy crisis. It goes without saying that this well-balanced approach to managing population pressure necessarily involves the use of incentives, coupled with the tactical use of mainstream disincentives.

- **Third and Fourth Worlds**

The developing nations of the Third World (see Fig. 30) are situated primarily in the southern hemisphere – for example, in South America, Africa, the Greater Middle East, the Indian Sub-Continent and Southeast Asia. Many have experienced first-hand the decline of imperialism throughout the postwar era. So, while Britain and France became increasingly preoccupied by European politics, the Soviets and Americans fought hard for the hearts and minds of these wider (often pre-industrial) societies, and in a manner which many would interpret as neocolonialism. These newly independent nations, however, remained largely politically non-aligned or neutral, i.e. they were neither agents of the communist Left, nor allies of the political Centre. Some, including Ho Chi Minh, probably perceived themselves as being principally **nationalists** (i.e. primarily concerned with developing their own national identity and affiliating only to that end).

A side affect of this comparative neutrality, which not even covert American-backed **CIA** initiatives could break, was a general lack of commitment to the future development of the Third World on the part of the two superpowers. Those developing nations which benefited the most, such as the Four Asian Tigers and Saudi Arabia, put away any anti-imperial grudges and forged strong links with the capitalist West. As for the rest of the developing world, seeking an autonomous path in an age of aggressively competing ideologies, neither superpower took to investing excessively in these rather erstwhile regions, save and except where it served their own narrow strategic interests. This more or less sealed the fate of some of the poorest nations on earth, save and except where naturally-occurring resources proved an irresistible lure.

With NATO membership defining the limits of the free world, these developing nations, who are rightfully suspicious of neocolonialism, have the opportunity to subordinate themselves to a neuro-cognitive presence which has already subordinated the UK, USA, EU and UN. This would, of course, be the most effective way to develop their country's social, political and economic infrastructure, sufficient to bring about the development they've so desperately sought. And all they have to do is avoid being non-aligned or neutral as regards female emancipation, freedom of speech and well-balanced politics. This is globalization with conditions attached – conditions which guarantee people's rights and freedoms globally, both now and in the future.

The term **Fourth World** has been used to describe those ethnic groups and cultures which, in many cases, are indigenous to a given region and who find themselves excluded from matters of state. Such individuals are forced to subsist, more or less on the periphery of modern political life, often due to the actions of their own governments. The term Fourth World also refers to the poorest and least developed nations of the Third World. Actively encouraging governments to participate in global progress, in the way proffered, would serve to raise the status of such groups, with or without the benefit of individual nationhood.

Ch15: Criminal justice

• Due process and the war on terror

The problem facing criminologists the world over is that the law, at its simplest, is simply a set of enforceable rules. Not all of those rules stand the test of time, and in the time it takes for society to advance some people will have been criminalized; as happened with the suffragette Dame Christabel Pankhurst (1880-1958) and the civil rights activist Martin Luther King Jr (1929-68). The suffragette movement in Great Britain and the civil rights movement in America, to give just two examples, would have been easily silenced by destructive forms of neuro-cognitive interference. Far better that these issues ran their course in the mainstream, rather than having their didactic value diluted by unseen forces. This is the chief reason why the phenomenon and its replacement shouldn't be part of the process of punishment – precisely because it would lead to an unhealthy heteronomous morality.

It would be easier to subsume the whole of mainstream society within a global presence with a limited remit, rather than one with a punitive manifesto. Besides, criminal justice is a domestic issue and therefore the sum-total of a given state's autonomous morality. Unquestionably, within a democratic society, the population is instrumental in determining what that autonomous morality actually amounts to. This is where we arrive at a healthy competition between the European Union and America, one which sees them compete to be the earth's most legitimate mainstream power. In America, that would amount to individuals honouring their constitution, which many have been forced to do ever since the Montgomery Bus Boycott in the 1950s, and the introduction of civil rights legislation in the 1960s.

The Montgomery Bus Boycott arose as a result of racial segregation on buses in the US city of Montgomery, Alabama. This boycott of the local bus service by African Americans was led by Martin Luther King Jr, who was arrested in the course of events and ended up spending two weeks in jail. King's own arrest for civil disobedience followed the arrests of several others, which came about due to black passengers refusing to surrender their seats to white passengers in compliance with local segregation laws. Four aggrieved African American passengers then seized the initiative and filed a lawsuit contesting these segregation laws. The resulting Browder vs Gayle case was fought all the way up to the US Supreme Court, who upheld the District Court's ruling that racial segregation on buses was unconstitutional.

Thus, the lines governing acceptable behaviour were re-drawn across America at the highest possible level, with the final arbiter being non other than the US Constitution itself. Further activism led to the passing of the **Civil Rights Act** (1964) and the Voting Rights Act (1965); with Martin Luther King becoming one of the truly outstanding figures of the 20th century (winning the Nobel Peace Prize in the process). This turnaround in the fortunes of black Americans was, in part, a reflection of **due process**; with the US Supreme Court ensuring that justice was administered in a fair and reasonable manner. It's fair to say that although under pressure from remotely-induced effects, Martin Luther King's actions and behaviour remained reassuringly authentic.

Contemporaneous with the civil rights movement was the rise of **Black Power** in America – and, in particular, the Black Panther movement. Founded by two black students, the Panthers began openly bearing arms and confronting what they perceived as white oppression. Martin Luther King incisively remarked that not only did white men have more guns, they were actually prepared to use them. This particular experiment in student radicalism and revolutionary violence ended with mass arrests, imprisonment and self-imposed exile. Whereupon, the whole movement foundered.

Simultaneously, on the other side of the Atlantic, white men were indeed procuring guns with the express purpose of using them in a violent manner, and not simply as symbols of power, i.e. terrorist groups were beginning to form, such as ETA, Red Brigade, the Baader-Meinhof Gang and the Red Army Faction, all of whom sought to bring revolutionary subversion to the heart of the new Europe. Ultimately, Martin Luther King's non-violent defence of the American Constitution had a high probability of success; unlike these unconstitutional acts of European terrorist violence, which simply afforded those responsible an uncertain status in the eyes of the law.

Concurrent with this outpouring of rage on the European continent was the dramatic reappearance of the IRA in Northern Ireland, resulting in a whole series of terrorist atrocities. What followed was the **Prevention of Terrorism Acts**, introduced in the UK between 1974 and 1989. These Acts were seen principally as emergency or temporary measures, necessary to prevent a short-term escalation in violence. Taken in conjunction with the subsequent peace process, begun under John Major in the 1990s, the results have been very positive indeed. However, by the new millennium far more wide-ranging anti-terrorist laws began to be introduced, aimed at any person who might be perceived as seditious or subversive. Thus, the Terrorism Act 2000 became the first of a new wave of statutes in the UK, but was quickly overtaken by events, as Act followed Act in the wake of 9/11.

Highly controversial, these statutes are often perceived as compromising due process, civil liberties and the rule of law. In essence, the state generally affords those accused of terrorist activities fewer rights than the average defendant, with the presumption being one of guilt, rather than innocence. This presumption of guilt, rather than innocence, has parallels with favouring punishment, and therefore constitutes an unruly use of power. More controversial is the issue of whether the general public should experience a loss of civil liberties due to palpably avoidable deficiencies in government. Expressed more bluntly, if a president fails to anticipate a terrorist attack should your civil liberties be sacrificed, as opposed to simply installing a more vigilant president?

In the wake of 9/11, or, if you prefer, following the failure of mainstream politicians to anticipate the very real danger posed by al-Qaeda, the only practical solution to mounting tension in the Greater Middle East was war on the ground in foreign climes. The Russians, as you will recall, had withdrawn from Afghanistan in 1989, and into that psychologically disturbed political vacuum there emerged an Islamic revolutionary movement termed the Taliban. The Taliban, who assumed control of Afghanistan in 1996, proved supportive of the al-Qaeda terrorist network, which was able to establish training camps in the country. Then, in October 2001, in the aftermath of 9/11, the US launched a full-scale invasion of Afghanistan to quell this growing menace.

Overwhelmingly successful, the Afghan invasion left the Americans controversially in possession of hundreds of alleged al-Qaeda members. These captives were then transported to the **Guantanamo Bay** Detention Centre in Cuba to await their fate. This detention centre presently consists of several camps (the original, intriguingly named Camp X-Ray, having closed). Almost a decade later, in April 2011, **Wikileaks** began publishing information regarding the alleged maltreatment of prisoners held in these camps, amid suggestions that their detention could be indefinite. Consequently, Guantanamo Bay has become synonymous with backpedaling on the issue of due process in those cases involving terror suspects. But of course, we mustn't forget, without due process there'd be no President Barrack Obama.

- ## The action potential of justice

The irrefutable abuse of Iraqi prisoners by the US military at Iraq's **Abu Ghraib Prison**, following the end of the Second Gulf War, contrasts sharply with the allegedly better treatment of terror suspects held at Guantanamo Bay. That contrast in treatment makes the issues surrounding Guantanamo Bay more a question of the suspect's legal status, if the official accounts are to be believed. Central to this issue, and indeed the whole workings of the criminal justice system, is the question of perception. A key feature of which is the wider conceptual framework, including news journalism, television reporting and the internet. Hence the threat posed by Wikileaks founder, Julian Assange (1971-), and the likely effect of his published information on the hearts and minds of the electorate. The challenge for the courts, whether military, civilian or otherwise, is to focus on the facts, given this highly-charged political atmosphere.

In essence, the law addresses allegations received from various sources, which has parallels with a person presenting science with a working hypothesis. If science, or the law, adopts your hypothetical assertion, they should nonetheless remain fully open-minded about the result, otherwise they'll be accused of subjective bias. Free of such partiality, the original assertion can then be tested through inquiry, and if repudiated the suspect is free to go, or, the hypothesis is abandoned; otherwise the evidence can be put to an independent scientific panel, judge, jury, or governing body, for their final verdict. The most common complaint therefore levied at the criminal justice system is the accusation of subjective bias. Bias which prejudices the outcome of the enquiry, and which, if present in science, would invalidate the result.

Therefore, whatever criteria you adopt for choosing a given assertion, the investigation itself should always be impartial; hence the need for objective third-party scrutiny – whether at the local level, or at the level of Guantanamo Bay detention. In the case of Guantanamo Bay, that scrutiny could well arise through the actions of the US Supreme Court – or, failing that, international pressure and a supranational presence judging events against privileged knowledge. This hypothetico-deductive approach, when applied to the law, clearly involves the adoption of a hypothetical position; which might very well conform to a given constabulary's expectations – but, given that not all those allegations are strictly falsifiable, science and the law remain markedly different devices. At its simplest then, an assertion is either repudiated outright, is dropped due to insufficient evidence, or, what evidence there is, is presented to a court for its deliberation – whereupon the court may convict, much depends on the weight of that evidence. This book borrows from both science and the law, in-so-far as it adopts a hypothetical position. However, an unbiased repudiation of this book's broader arguments remains elusive, due to some rather paradoxical politics (brought about by the demands of continued deception). As for the law, it possesses the equivalent of an action potential, not unlike the one found in the membrane of a neuron. If unbiased police inquiry produces a statistically significant result, then a case is actioned, i.e. the case is passed to the Courts for their consideration. Otherwise, the case is dropped and inactivation occurs, i.e. a return to normality.

What shouldn't happen is the judicial system entering into some protracted semi-excited state without adequate resolution – which would, of course, prove draining for all parties. This is the danger in the Julian Assange saga, which is presently running its course at the time of writing, with the said Wikileaks founder yet to be formerly charged in relation to various alleged sexual offences (allegations which Assange strongly denies, and which are said to be politically motivated). NEURON would be party to privileged knowledge, making that saga a test of logic, wisdom and integrity for all involved.

The presumption of guilt puts the onus on defending parties to rebut the hypothesis of wrongdoing or face incarceration. Besides actively prejudicing any inquiry and rendering the law apathetic, it may be impossible for genuinely innocent parties to prove innocence. This why the **presumption of innocence** is the correct one. The abandonment of due process and the presumption of guilt leads inexorably to simple denunciation (common to both fascism and communism). Consequently, it plays into the hands of the political extremes and elective dictatorships. It was through a process of routine denunciation that the Third Reich deported, incarcerated and murdered many of its communist opponents and outspoken critics. Therefore, we abandon these principles at our peril – principles which serve to define where the political Centre actually lies.

Where a case is committed to open court, there is a stratified judicial system to pronounce on the same. Those laws which are alleged to have been broken having been enacted by the legislature with the support of the executive (**statute law**), or introduced by the judiciary directly (**common law**). Stratification within the judicial system, both at home and abroad, reflects the relative gravity of the case in question. Hence, the court system in England and Wales, for example, comprises County Courts, which deal with civil cases, (e.g. bankruptcy, divorce and personal injury claims), together with higher courts, namely, Magistrates Courts and Crown Courts, which deal with criminal cases (e.g. theft, fraud and murder). As if taking its lead from the US, these Courts now come under the overall jurisdiction of a UK Supreme Court.

Across the Atlantic, in the US, Congress has divided the United States into 94 Federal Judicial Districts, every one of which contains a District Court (which hear both civil and criminal cases). Additionally, America has several specialist courts, with all US Courts (specialist or otherwise) coming under the jurisdiction of the US Supreme Court. As for the various states which make up the European Union, they have their own domestic Courts which are subordinate, in theory, to both the **European Court of Justice** (which adjudicates in respect of alleged breaches of EU law), and the **European Court of Human Rights** (which upholds the right of free expression in the EU; whilst safeguarding against discrimination and human rights abuses).

Europe is also home to the world's foremost International Court; namely, the United Nations very own International Court of Justice, which sits at The Hague in the Netherlands. Recently, The Hague has established a number of specialist courts in the aftermath of crimes against humanity (up to and including the crime of genocide); for example, the International Criminal Tribunal for Rwanda and the International Criminal Tribunal for the Former Yugoslavia. Occasionally, a country's very own legal system is brought into play, as in Iraq. Here, the Supreme Iraqi Criminal Tribunal presides over issues relating to human rights abuses, war crimes and genocide, committed between 1968 and 2003, during the rule of the Ba'ath Party (prior to Saddam Hussein's deposition in the Second Gulf War). This is the tribunal which sentenced Saddam Hussein to death by hanging in 2006.

It's worth adding, that in the case of Osama bin Laden, closure came in the form of death without trial, due to shots fired by US Navy SEALs in the course of a gun-battle in Pakistan, in 2011. Taken together, all of the above examples, from the simplest and least controversial to summary justice in the course of military action, all beg questions relating to **jurisprudence** – in other words, the theory, science and philosophy of the law.

• Normative jurisprudence

Originally what law existed was simply a reflection of commonly held values and beliefs, in other words **natural law** (i.e. the law is no more than a reflection of morality). As societies advanced the law became heavily laden with legal banalities, related to such disparate matters as copyright, contracts and conveyancing, and so the term **legal positivism** gained currency (i.e. the law is simply what is posited, often for rather utilitarian reasons). Later, the growing infrastructure of the law, and the bureaucracy surrounding the same, began to define the very nature of the law itself, leading to what is termed **legal realism** (i.e. the law is simply what arises, or is likely to arise, given a whole array of varying factors, some related to income and geography).

At the heart of contemporary judicial proceedings one finds a strong element of natural law, allied to the key principles of **natural justice** (an aspect of which states that a verdict cannot stand unless the person directly affected was given a fair opportunity to state their case and to know and answer the other side's case). Because natural law corresponds with the kinds of activity which society expects to trigger revulsion and disquiet, one would expect the state to address those issues rigorously and at its own expense. Beyond such revulsion and disquiet, however, the terms legal positivism and legal realism become more pertinent, with the law becoming nothing more than a set of convenient rules, backed up by the means of enforcement.

Clearly an overarching supranational presence can evaluate the performance of mainstream criminal justice systems against privileged knowledge, rewarding those elements which appear to excel. By this means future justice is more accurately administered in the mainstream, with judicious individuals gaining certain telepathic advantages. In keeping with the concept known as *habeas corpus* (Latin: *have the body*) NEURON would surrender the defendant to the mainstream, should punishment be deemed necessary. Natural justice dictates that not only does the defendant have the right to know and answer the other side's case, they also have the right to be punished by conventional means and in a manner which can be subject to appeals and legal challenges. In this way the electorate remains fully accountable for the balance it chooses to strike through its elected representatives.

The alternative would be a phenomenon extra-judicially killing in a summary fashion, without recourse to natural justice and in a manner which absolves the wider population of culpability. To avoid this wholesale diffusion of responsibility (not to mention mass murder) we separate power, make the law a mainstream psychometric test of logic and integrity, and allow democracy to determine the precise limits. Moreover, a person's conative dimension, by which we mean their overtly expressed attitudes, impulses and desires is easier to determine if they aren't burdened with radiological ill-effects. Scapegoating is also a potential hazard, whereby an individual professes abuse, which is wrongly attributed to mainstream peers and authority figures, but where the actual abuse stemmed from remotely-induced effects by unknown third parties.

Ultimately, the law should favour rewarding a given defendant with their freedom if the presumption of innocence cannot be fairly rebutted. It should also endeavor to use punishment sparingly and only to define the limits – using non-custodial sentences where possible, and making the death sentence available only to The Hague. NEURON, for its part, would judge all persons in relation to the law (and, indeed, the law itself), with those who truly excel being inducted into this supranational normative device.

· **Commission, omission and possession**

No matter what the law is posited as being there normally has to be two essential ingredients for a crime to have been committed; namely, **actus reus** (Latin: the guilty act) and **mens rea** (Latin: the guilty mind).[52] NEURON would have definitive proof of thinking, but no mandate to punish; whereas the mainstream would have the prerogative of punishing, but no such proof. By this means the world's judicial systems are prevented from punishing thoughts, as cognition cannot be proven. One would even go so far as to prohibit the act of punishing thought – making those who try, guilty of a felony. Additionally, employment law would interpret dismissal, on the grounds of cognition, as unfair. The net effect of these provisos is to maintain the criminal justice system in much the same form, whilst providing for its improvement.

The only tangible impact of a paradigm shift would be the sense that one's thoughts are available to those with the power to selectively advantage. In fact, the science of what is legally knowable at a mainstream level, in an age of growing telepathic awareness, might be termed **epistemological jurisprudence**.

Such experts would be concerned with determining the foundations, limits and reliability of knowledge, as regards the law, given the whole new paradigm which would then exist. Provided that such effects are properly applied, their presence needn't be raised in court. However, a general defence of having been subject to ill-treatment shouldn't be ruled out (obviously expressed using apposite Latin phraseology).

A criminal misdemeanor entails committing something, omitting to do something, or possessing something (respectively, **commission, omission** and **possession**). Whether one is dealing with commission, omission or possession, the guilty act (*actus reus*) and the guilty mind (*mens rea*) would still need to be proven to the satisfaction of the court. Except, that is, in the case of **strict liability**, where only *actus reus* needs to be proven. Assuming, for the sake of argument, that those applying the effects aren't strictly liable, then if they are seeking to raise awareness, whilst educating society as to the implications of such effects, and actively promoting women in the process, then *mens rea* may not apply – even if the acts themselves appeared *prima facie* criminal. According to this scenario no crime has been committed.

If that excuses those applying such effects, what of those to whom such effects have been applied – surely they deserve an equally magnanimous interpretation of the law. The defence of duress would stand if the phenomenon had employed physical violence, grossly interfered with physiological processes and/ or engaged in excessive ill-effects, sufficient to render the individual diminished in their judgement. With the phenomenon suitably revised, arguments such as these are likely be confined to historical claims; but claims which may have increased weight and importance if tangible proof of overall good-intentions doesn't surface shortly.[53]

A constructive paradigm shift, together with the timely assimilation of neuro-cognitive powers, would do much to rebut the assertion of criminal intent. Making their application nominally legal – with *mens rea* very much the exception, rather than the rule.

•   Cruel and unusual punishments

Without doubt, the mainstream is unlikely to approve a neuro-cognitive presence if it conflicts with the European Convention on Human Rights, the UN's Convention Against Torture (and other Cruel, Inhuman or Degrading Treatments and Punishments), or simply America's 8th Amendment (which actively forbids cruel and unusual punishments). Thus, public authorities have a limited but positive duty to protect people from harmful interference and destructive ill-effects. This book proposes that although inhuman and degrading treatment is potentially possible, this shouldn't deter us from assimilating these effects into our daily lives. Crucially, we need to examine the legal framework surrounding cognition extremely carefully, so as to gift the unborn complete freedom of thought – and, by definition, make all things possible, save and except those things prevented by the people themselves.

The UK's Police and Criminal Evidence Act allows the police to arrest someone, without a warrant, who they have reasonable suspicion is about to commit an offence. One can only imagine the potential chaos which might ensue if the police presumed to regulate mainstream thinking. Arresting someone on the basis of their thoughts, in the absence of proof, would amount to **false imprisonment**. This in itself would classify as a **tort** (i.e. a wrongful, unwarranted and illegal act for which one might be fairly compensated). The person normally answerable is the one who committed the tort, but **vicarious liability** would apply in this instance, in which case the constabulary or police department would incur an action. This all serves to protect the interests of the general public, who aspire to think as freely in the future as they do today.

Some would argue that if thought and cognition are known, crime should be actively prevented from happening. But that presupposes that a supranational body should intrude on every single aspect of home affairs. It also implies that domestic politics isn't a mature test of societies, cultures and people. If a truly democratic society permits firearms to proliferate, allows grinding poverty to appear, and enriches people's perceptions of themselves and each other with arbitrary acts of dramatized violence, then the ramifications of those choices will play-out authentically in the mainstream. Equally, civil rights activism may be interpreted by the state as criminal, but not by an overarching presence, which prefers that such crimes be committed.

Ultimately, power should concentrate in legitimate hands, which is far more important than zero tolerance. This is due to the nature of power in the postmodern age, which has a ferocity, magnitude and profundity so great, it cannot be permitted to find its way into the wrong hands – namely, those ill-balanced individuals who weaken the core principles of natural justice, i.e. *audi alteram partem* (no one should be condemned unheard) and *nemo judex in causa sua* (no one should be permitted to bias legal proceedings).

The state, when viewed as a person in the eyes of international law, may be guilty of committing a crime, omitting to act, and/or possessing, for example, weapons of mass destruction. The view that one can legitimately apprehend all one reasonably suspects are about to commit a crime, when applied to nations, leads ineluctably to the invasion of potential belligerents. Crime statistics suggest that men are more likely to externalize their issues than women, with male-dominated one-party systems invariably producing history's foremost aggressors. But the grounds for taking action against rogue states must be considerable. Any action taken against such a state, based solely on the cognition of its nationals, would be illegal. This serves to protect all, including allies, from having to deal with the vagaries of their own psychology.

Legally-speaking, one can't simply conjure-up a transcript in court and assert that a given telephone conversation took place – one would need some form of written proof from the service provider that such a call occurred. This proviso obviously doesn't prevent two people from engaging via such radiofrequency devices, but it does place limits on how far they can go in proving the details of that activity in court. Transcripts of a person's thinking, presented to a court without corroborative evidence from NEURON, would be deemed wholly **inadmissible**. Moreover, NEURON would be constitutionally obliged not to surrender such evidence. This protects against defamation; as those who can't prove thinking to a court risk litigation when alluding to the same.

By these means one can think as freely as one chooses and without the risk of false imprisonment, military action or defamation. However, one can't be complacent about the effects themselves, as their unruly application may all too easily pervert the course of justice. A defendant stood in court today might consider a fair trial impossible, given the wholesale suppression of corroborative information regarding their maltreatment – ill-treatment which wouldn't arise in the phenomenon's reformed state. Additionally, this presence has the capacity to severely agitate, making not just crime but litigation itself far more likely. Therefore people, societies and cultures need to be assured this isn't to be an *agent provocateur* transcending national boundaries.[54]

- International law

As we've seen, the medium of the brain has remained largely unchanged, and will remain unaltered if we simply obscure the truth regarding the same. Accordingly, the mainstream is a rich milieu, transparent to a select minority, within which disparate forces play-out, sometimes uncomfortably. Every state, by definition, possesses a permanent population, viable government, and a defined territory over which its courts have legal jurisdiction. Above and beyond that, however, international law applies. **International law** encompasses land, sea, space, and even aliens (or, should I say, displaced persons; often subdivided into resident aliens, transient aliens and enemy aliens). Such law comprises so many customs, treaties and conventions; but crucially no international legislature, as such. Even the UN's own International Court of Justice operates solely on the basis of consent – making the mainstream seem a little unruly on occasions.

That over-reliance on consent is never more apparent than in the case of the Nuclear Non-Proliferation Treaty, which is consciously flouted by several countries. Thus, international law is posited, often for very well-intentioned purposes, but what effect that has, realistically-speaking, varies according to the personalities of the states themselves. Even when faced with crimes abhorrent enough to defile natural law, the international community can appear faint-hearted, heavily-divided and weak. So, perhaps an overarching extrasensory power, selectively advantaging certain elements, would break the impasse. After all, the rewards for acting decisively in the global interest might serve you in the post-peak future.

At least as serious as resource scarcity, nuclear weapons proliferation and global warming is the question of women's rights worldwide – rights which remain vulnerable to political inertia and backpedaling in far too many countries. A neuro-cognitive power, of the kind described in this book, would hand women an unprecedented global advantage. That advantage wouldn't involve cruel and unusual punishments, but rather the highly selective advantaging of those with merit. Additionally, because this presence would be devoid of *mens rea*, or the guilty mind, it would transcend the whole of international law. What then subsequently happens, as they say, happens.

PART VI

Future

Chapters 16-18

Ch16: **Believers**

• Thanatos and the Third World War

Psychoanalyst Sigmund Freud (1856-1939) perceived, within the human psyche, an immensely destructive potential – one so bound-up with negative energy and unruly emotions that he termed it the death drive. Later afforded the name **thanatos**, the death drive represented an uncompromising death instinct, obdurately expressing itself at all levels. The gross externalization of this inner force would, in Freud's view, draw the world into so much barbarity and slaughter. These forces were seen by Freudians as being the very antithesis of the life-affirming energy, known as **libido**. So, the human condition appeared to embody both the need for sex, pleasure, comfort and security, together with the apparent need for self-defeating aggression, violence and the pursuit of suffering, often for its own sake.

These rather Paleolithic drives appeared to echo the minds of those venturing into northern climes, in our ancestral past. People who were seeking to maximize the available rewards, whilst reducing, as far as possible, the more punishing effects of climate and geography. It's possible that altruistic self-sacrifice, articulating itself as a lack of concern for personal safety, played a significant part in the overall evolution of society. That what the individual lost through an uncompromising death instinct, society gained through the rewards of eventual success. Clans, tribes, societies, and later nation-states, became the perfect receptacles of this boundless energy, with the obvious potential for wider conflagration. Such energy was never greater than that witnessed in Germany in the early 20th century.

That heavily suppressed energy, much of it expressed as military muscle, concentrated in the hands of the German Fuhrer Adolf Hitler and his Axis Allies, whose gross externalization of their respective grievances drew the rest of the world into the unholy quagmire of the Second World War. Thus, the rich tapestry of nations, each with its own personality and psychology, demands conscientious management from above; but not, it must be said, repression. After all, the mainstream must be understood for the right people to be inducted into NEURON, and for the wrong people to be distanced from the same. So, as the English-speaking world began to harness telepathically-induced effects, the forces of the extreme Right and radical Left began to violently coalesce; with the psychology of their constituent parts coming, as they say, into full view.

In his book, The Search for Peace, former UK Foreign Secretary Douglas Hurd (1930-) argued that if Britain and France had gone to war with Nazi Germany when Hitler first occupied the Rhineland, in 1936, the Allies would have won (since German rearmament was still in its infancy).[55] Had they done so, would that have been the proffered death drive? Some might argue that failure to take that initiative was the death instinct negatively expressed, i.e. wholesale acquiescence as regards the breaking of the terms of the Treaty of Versailles brought with it even greater death and destruction across Europe. A case of death drive if you do, death drive if you don't!

Several generations later, in 2003, an American-led coalition removed Saddam Hussein from power – a person described by the Bush Administration as the new Hitler. But if Saddam was the new Hitler, Iraq was certainly no Germany. If Britain and France had invaded Germany, in 1936, it seems highly unlikely that they would have been as successful at nation-building in the wake of occupation. According to this synopsis long-term peace and stability across the whole of Europe was easier to arrive at (back in 1936) through military conflagration, tragic though those events were, rather than premature occupation. Much depends on whether events reflect the personality of the leader, or the overall character of the nation. In the case of the former, the answer is occupation; in the case of the latter, regrettably, the answer is conflagration.

Behind the façade of thanatos, therefore, lie much deeper questions of strategy and the long-term pursuit of peace and security in the world. It was better to enter into war with Germany prior to its development of the atomic bomb, and the consequences of not doing so remain unthinkable. The writer isn't convinced that the two world wars were simply grandiose expressions of self-defeating **thanatic** violence, but in reality something decidedly more profound. As such, we can't assume they won't repeat themselves. But if they do, a supranational presence will make believers of all those empowered by its presence, as they overcome the unruly forces ranged against them. At the conclusion of hostilities balanced-mindedness will have prevailed, with the victors appearing strongly telepathic. Telepathy and balanced-mindedness becoming forever intertwined.

199

• Peak religion and its secular aftermath

It's possible that we are witnessing peak faith, in which the production of religious dogma has reached its apogee, beyond which we can expect to see a steady decline and eventual exhaustion of orthodox belief.[56] This argument suggests that atheism is likely to become more prevalent in the future, with secularism displacing theism in the world's most advanced nations. **Atheism** being the antonym of **theism** (the former believing that there is no supreme deity and the latter believing that there is). As faith continues to play an important part in many people's lives, informing the conceptual framework which colours the perception of both friend and foe, it cannot be easily dismissed from this rather expansive synopsis. Objectively-speaking, there is no God – or, if there is, they're not omnipotent, or choose not to appear so.

Those who insist there is a God are faced with a dilemma; in-so-far as this supernatural presence must combine a global reach with a strictly limited remit. Having found the position of God apparently vacant, and in deference to the invocations of millions, we arrive methodically at the next best thing – namely, the personification of women's best instincts played-out at the grandest of scales, via radiofrequency effects. In the final analysis, is a God that's constitutionally incapable of being anything other than self-limiting intrinsically stronger than the neuro-cognitive power one proposes? Done badly, we have the means to create a devil in our lives. But don't expect a naturally self-limiting God to be *more* powerful than the devil we create.

The writer would rather that your well-intentioned prayers didn't fall on deaf ears; but rest assured, they will, if we don't construct something more sublime. Provided the phenomenon is suitably revised and its replacement inspires widespread trust, then a **secular belief system** may arise. Such a belief system would obviously facilitate the more effective use of extrasensory experiences, as people learn to trust to their perception. Who'd have imagined that the *Aufbauprinzip* (or building-up principle), at the heart of all modern chemistry, would lead to a naturally self-limiting creator, whose inability to punish makes it all the more mysterious. Provided all those who are inducted into this **supranatural** presence are free of the worst excesses of thanatos and libido, then the overall net-result should remain positive.

Unfortunately, the phenomenon which arose during the course of the 20th century has deviated, exhibiting signs and symptoms consistent with it favouring punishment; using reward selfishly, and having no limits to speak of. It's inconceivable that we'd deliberately create a devil in our lives – more probable, it's a symptom of inadequate self-regulation, coupled with chauvinistic excess. With the incorporation of far more women, sufficient to imbue it with the soundest femininity, it might grow in strength. A strength which the Greater Middle East, in its present patriarchal form, can only pray for.

- ## The Greater Middle East and the Arab Spring

The Greater Middle East comprises a huge swathe of overwhelmingly Muslim countries, all lying on, or very close to, the line of latitude which is 30^0 north of the equator. If one follows this line from west to east one passes straight over Morocco, Algeria, Libya, Egypt, Israel, Jordan, Saudi Arabia, Iraq, Kuwait, Iran, Afghanistan, and finally Pakistan. Additionally, there are several states to the north; principally, Tunisia, Lebanon, Syria, Turkey and some Muslim-dominated former Soviet republics. To the south of this line one finds Sudan, Yemen, Oman, United Arab Emirates and Qatar. The Arab Spring, which began in December 2010, has witnessed rulers forced from power across this dustiest of swathes. Today, through a combination of mass protests, spontaneous rallies and civil uprisings (many aided by the shrewd use of social media and the internet), the whole region has entered into a state of much-needed flux.

Whereas the 20th century witnessed the crucible of the entire world surrendering its psychology to critical scrutiny, in the present century that emphasis has shifted towards specific regions. For the Greater Middle East's current history to be properly understood, and for the right people to be inducted and distanced, events must authentically run their course. That will backlight those who are the region's future, and who will gain most from heightened awareness. Such authenticity is undoubtedly the key to global peace and stability, provided an overarching presence rewards those with sufficient merit. Without the benefit of such rewards, authenticity, and even goodness, become potentially superfluous or academic. This explains humanity's willingness to conjure-up divinities. It's because humans pray that something will make the kind of value judgments that nature avoids making in their lifetimes, but logic dictates it should.

Having a supranatural presence which is rewarding in its construct therefore avoids a blind and selfish determinism manifesting itself. If the Greater Middle East chooses to appropriate the political apparatus of the free world, history will no longer be suspended. If it then provides for remotely-induced effects, the average citizen might subsequently profit from that ensuing history. At present, the set of readily enforceable rules imposed by Islam has tended to penalize behaviour we take for granted in the free-world, making felons of those who might otherwise be valued as writers, artists and free-thinkers. Tunisian street vendor, and **Sakharov Prize** winner, Mohamed Bouazizi (1984-2011), whose self-immolation catalyzed the Arab Spring, didn't externalize his inner pain in a way which harmed others. In fact, his spirit was such that within NATO's direct sphere of influence he would have been more fairly rewarded in life.

- NATO's model empire

The free world comprises the US, UK and EU (together with all their allies); with NATO very much its strategic backbone. Within this arrangement one finds no less than three recognized nuclear states. But, whilst this, the First World, has gone from strength to strength, the Second World has experienced endemic conflict, fragmentation and ideological meltdown. First came the Tito-Stalin split, then the Sino-Soviet split, then the wholesale dissolution of the USSR, followed by the death of Yugoslavia, growing ideological uncertainty in China, and the 'big-bang' expansion of the EU (via the Treaty of Athens). That consolidation of power and influence in the West comes not from nations cleaving loyally to the USA, but from American science engineering a supranational presence which transcends the USA's own domestic politics.

As for the UK, if it allies itself to an America compromising due process, it may find that the English-speaking world begins to fall behind the European Union. If the European Union and America engage in a healthy competition for power and influence, the UK may find that it can afford to abandon itself to a federal Europe committed to being the most legitimate power on the face of the earth. Alternatively, the UK may strive to remain largely independent; with America the head, the UK the heart, and Europe the body, of an indefatigable global alliance reaching out to all four corners of the globe. Such a pick-and-mix "transformers-style" empire combines social justice, rule of law, and healthy competition, in ways which serve humanity wisely. The UK needn't choose between Europe and America, it could remain precisely where it is, at the heart of NATO. [57]

However, an overarching neuro-cognitive presence is unlikely to concentrate undue power and influence in the hands of such a geopolitic, whose mainstream aspirations cannot hope to match those of either Europe or America in the long term. In answer to the question what should the US, UK and EU actually do, the answer is: whatever serves the free world.[58] For example, by mutual consent the UK could remain precisely where it is, Europe could expand, and the three could corner the world's markets in a manner which serves the global economy. Additionally, this growing NATO alliance could help to make the peace in various places. To understand how, lets examine some of the world's most troublesome hotspots – starting with the Caucasus.

Abkhazia and South Ossetia are two republics which border the Russian Federation and Georgia (see Fig. 31).[59] Russia recognizes Abkhazia and South Ossetia as states in their own right, whereas Georgia refuses to. Georgia is presently seeking membership of NATO, and wants to incorporate Abkhazia and South Ossetia into a unified Georgian State (albeit with a certain amount of autonomy). This is an example in which NATO expansion, through incorporation of Georgia, could lead to conflict with Russia. But in many ways the free-world is best served by an amalgamation of Russia. Engineering a paradigm shift might make possible an enduring diplomatic solution. How well these republics subsequently co-exist might then be down to external pressures related to energy and mineral security – after all, these regions contain not just valuable resources, but also the means to transit the same from farther afield.

Another republic facing comparable stresses is the Chechen Republic, whose Muslim factions fought to be independent of Russia (see Fig. 31). Since 2003, however, Chechnya's constitution has bound it squarely to the Russian Federation, albeit with a certain amount of autonomy. As mentioned previously, the year after this constitution came into force Muslim separatists stormed a school in Beslan (Beslan School Hostage Crisis, 2004), resulting in a siege which killed more than 300 hostages, many of them children. The Caucasus are therefore the Balkans of the 21st century, replete with so many fractious wannabe states, hot-headed seditious factions and competing global interests. Not that the Russian Federation would want to be drawn into a conflict with those openly professing an extrasensory potential and far-reaching powers.

Figure 31: NATO members and global partners

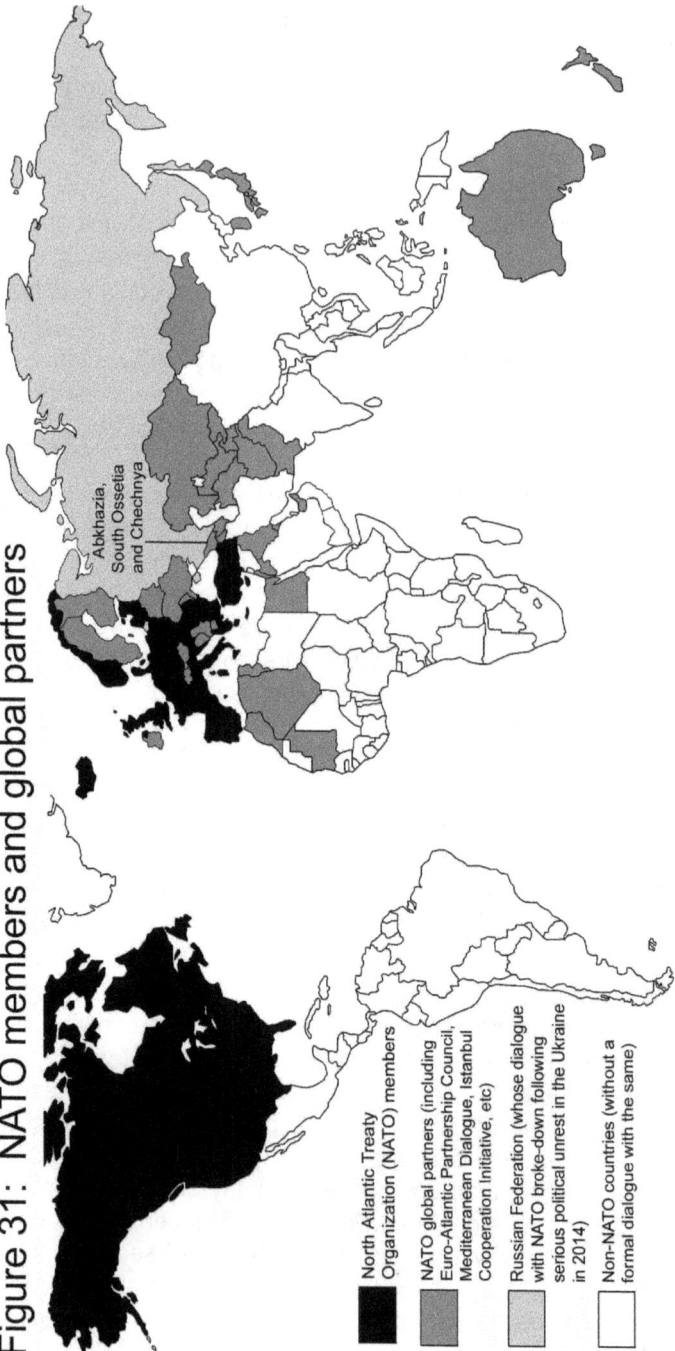

Abkhazia,
South Ossetia
and Chechnya

North Atlantic Treaty
Organization (NATO) members

NATO global partners (including
Euro-Atlantic Partnership Council,
Mediterranean Dialogue, Istanbul
Cooperation Initiative, etc)

Russian Federation (whose dialogue
with NATO broke-down following
serious political unrest in the Ukraine
in 2014)

Non-NATO countries (without a
formal dialogue with the same)

In addition to the tensions in the Caucasus, other seismically-significant political hotspots around the world include the standoff between North and South Korea; Iran and its nuclear program; Israel and its Arab neighbours; the question of Chinese governance; and continued unrest in the Ukraine. Only the Korean Peninsula and the question of Chinese governance lie well beyond NATO's present geographical limits, the others being more or less disputes along its borders – Turkey having joined NATO in 1952. However, if Russia merges with Europe, both Chinese governance and the Korean Peninsula become issues close to NATO's geographical limits. This arguably makes the Caucasus the most significant of all the world's hotspots, provided that Ukrainian separatism doesn't escalate.

It remains to be seen whether Abkhazia and South Ossetia drive a wedge between Russia and NATO, or whether Russia becomes a part of Europe. If Russia did become a member of the EU (and NATO); Abkhazians, South Ossetians and Chechens would have little choice but to become citizens of the First World, in an age when to be a citizen of the First World is to be subsumed within a telepathic milieu. Even Mongolia, as we've seen, has become a multi-party democracy following the sudden collapse of the Soviet Union. So, if NATO expands its borders, consuming no less than three-quarter's of the earth's defence spending in the process, it will become as expansive as it is unassailable.

- Sino-First World relations

Sino-First World relations are inherently complex. The Korean War tested the Second World's mettle in an age when the communist bloc was still relatively coherent, and it had the effect of showing precisely whose side the UN would be on in the face of communist aggression – unequivocally on the side of the free-world or Western Allies. The subsequent Sino-Soviet split (1960-1989) further amplified that imbalance of power in the West's favour, since which time the West has enjoyed a relatively stable relationship with mainland China. In fact, some writers have suggested a growing interdependence between the US and China, and that far from being strategic rivals they have the air of economic partners. The writer believes that in spite of its ongoing industrial revolution, China remains by far the more dependant of the two.

Having been a largely agrarian society, the communists who won the Chinese Civil War in the late 1940s demonstrated their contempt for the bourgeoisie and aristocracies of foreign capitalist economies, and indeed their own wealthier landowners, by mythologizing the proletarian cause. During the **Great Leap Forward** (1958-1961) tens of millions died through forced **collectivization** – conceived of as a means of providing for the growing cities. Next came the **Cultural Revolution** (1966-76), in which Mao Tse-tung (1893-1976) promoted his own inimitable interpretation of Marxist-Leninism, termed **Maoism**. Maoism basically amounted to a crude attempt at stamping-out grass-roots capitalism, often through the routine use of violence and intimidation.

In many ways George Orwell's Nineteen Eighty-Four could easily be adapted to reflect the life of a Chinese functionary during the Cultural Revolution – one eking out a life in one of Mao's experiments in urbanization. Public executions were common at that time, as was police brutality, abuse of opponents, and unprecedented levels of suicide. Millions are thought to have died during these counter-revolutionary purges and the widespread persecution of liberals. Amid this Orwellian nightmare, in the year Nineteen Sixty-Four, the Chinese exploded their very first atomic bomb, making them the world's fifth nuclear power. Having become a nuclear power the United Nations had little choice but to relent, allowing them to assume China's seat at the UN.

Hu Yaobang (1915-89), who was General Secretary of the Chinese Communist Party from 1982 to 1987, helped to dismantle the widespread cult of personality which had arisen around Chairman Mao. In doing so he acquired a number of political enemies within the party. Forced to resign by conservative elements, he ceased to take an active part in politics. Following Hu's death in 1989, well over 100,000 students descended on Tiananmen Square, in central Beijing, to demand political and economic reforms, during a whole series of demonstrations which looked set to gather momentum. Crucially, the government moved to end these mounting protests with the systematic use of violence; resulting in many thousands of student deaths, both in Tiananmen Square itself and as a consequence of copious arrests and executions. This more or less cemented the global view of China as an out-dated Cold War relic.

Such bullish adherence to Marxist-Leninist dogma has served to create the disagreeable persona we commonly associate with present day China. Trade relations with the rest of the world, including the West, are generally good, due to the widely-held belief that such relations will encourage a liberalization of

China's economy – and, in time, its politics. Nevertheless, China continues to provoke criticism and condemnation for its casual use of detention without trial, alleged use of torture, over-reliance on the death penalty, and scant regard for environmental concerns, particularly in relation to CO_2 emissions. These, however, are mainstream concerns, sent to tax the minds of those directly involved in overseas diplomacy, international relations and the careful crafting of foreign policy at the most conventional of levels.

Currently, China and Russia are the foremost members of the **Shanghai Cooperation Organization** (SCO); which, through its various affiliates and member states, is quite simply the nearest thing there is to NATO's opposite number. The SCO is said to account for half the human race, if some estimates are to be believed, but has no plans, as yet, to be seen as a **military bloc**. In any event, NATO, with its extrasensory capability firmly established, would be by far the more formidable. Moreover, the West's commitment to the legitimate use of power, together with its widespread promotion of democratic politics, and support for fundamental freedoms, would hopefully appeal to many within the SCO, not least of which, the Russian Federation itself.

If the strategy presented in this book works the next major phase of human progress should proceed smoothly enough. But failing that we may be faced with the potential for global conflagration. How long such a conflict would last it is impossible to say, but one can guarantee that the victors would emerge powerfully telepathic and supremely well-balanced. Eventually, whether through Third World War or simply the pacific restructuring of our lives, we will arrive at the global assimilation of these technologies. And, when we do, we will have entered the planetary phase of civilization. Far from being dehumanizing, this brave new world will be far less oppressive than many of the politics it ultimately supplanted.

- ## The planetary phase of civilization

Systems theory perceives the earth as a collection of discrete but nonetheless interconnected systems. Thus, one finds the meteorological system impacting on the political system, which contains numerous individuals possessing a broad range of physiological systems, not least of which, their neurological system. A feature common to all such systems is that they are subject to periodic change, adaptation and adjustment: be it seasonal variation, revolutionary upheaval or relentless genetic evolution. Science perceives these discrete entities as being complex adaptive systems, and their systematic study as so many branches, fields and disciplines. In fact, the persona's very own chemical imprint is the product of this complex conjuncture of disparate systems; systems which are themselves in a relative state of never ending flux.

In many respects the existence of specific branches of science confirms the existence of discrete systems: systems which can be either open or closed. How one defines open and closed, however, is very much a matter of interpretation. Predictably, physics stumps-up a hard-and-fast definition, which can be appropriated for the purpose of our wider debate on telepathy. In thermodynamics an **open system** is able to exchange energy, heat and mass with its surroundings; whereas a **closed system** can only exchange heat and energy. Moreover, an **isolated system** (which is said to be a theoretical impossibility in nature) cannot exchange anything with its surroundings.

At the molecular level, all of one's physical processes must be open systems for the preservation of life; in-so-far as one requires sustenance, in the form of food and water, to provide, for example, all the necessary carbon, nitrogen, hydrogen and oxygen, which together make-up 96.6% of the human body. By engineering large covalently bonded macro-molecules, surrounded by others of varying complexity, and all linked together by a variety of forces, the human body adds to the list of recordable spectra. The mind arises, therefore, due to person-specific absorption and emission spectra, with the brain preferring to exchange mass with its surroundings, rather than energy or radiation.

What might we term this arrangement, i.e. one in which mass is exchanged, but not energy? It appears that mother nature has confounded physics, by conjuring-up a state which science itself failed to anticipate. The molecular state providing for induced effects is therefore neither open, closed, nor isolated. One might term this arrangement a **semi-isolated system**, in which mass seeks to exchange little or no energy with its environment, only with itself.

The planetary phase of civilization therefore involves the mass-extinction of these semi-isolated phenomena, as technology prises open these partially isolated worlds (one of which happens to be your mind). This is every bit as powerful as splitting the atom, because tapping-into that part of ourselves which was never meant to be accessed could be our undoing as a species. Consequently, the 21st Century is as much about the conservation and preservation of our minds, as it is about population control and sound energy policies.

Ch17: Society and culture

- Rebel, artist, hero

Whatever else personality is, it must be authentic: making personality a veritable work of art. The art of being yourself, whilst under the influence of radiological effects, and then burning that image deep into the fabric of history, is the stuff of popular mythology. You are, quite simply, mother nature's best attempt at filling-in the gaps in the electromagnetic spectrum. Such exclusivity has real currency in free societies, because every mind is capable of exhibiting a never to be repeated pattern of thought, conscience, action and behaviour. The phenomenon which has emerged during the last one hundred years represents a global consciousness; providing for the deepest of introspection, the broadest of analyses and the most peripheral of all nervous systems. Both the brain and the societies within which radiofrequency technologies are rooted are absorbing and emitting within themselves, making western civilization neurological on a colossal scale.

Whilst Second World history became saturated with political propaganda, and Third World history with tales of despotism, overpopulation and the environment, the narratives associated with the First World have been far more absorbing. Beyond the First World the law has largely politicized activities falling outside some rather narrow social norms, whereas in the West a prudent revision of the absolute limits has resulted in self-expression becoming generally apolitical. Some aspects of this western culture have arisen in a mostly continuous fashion (for example, the gradual break-down of the class system and the emergence of a **generation gap**); whilst other aspects have materialized in a largely discontinuous manner (for example, the invention of the electric guitar and the mass-production of personal computers).

All of these are relatively hollow advances without the associated process of enrichment which invariably follows. Thus, the overall **corpus** of knowledge surrounding the electric guitar enriches every chord and amplifies every riff. At the heart of this brave new world lies the somewhat complex, prickly and mercurial human spirit; vulnerable to the dynamics of surveillance, ill-balanced feedback, attributional errors, and the wholesale destruction of one's inner-self. Somewhat audaciously, and against all the odds, a given individual might nonetheless assert themselves and add meaningfully to our collective perception. So, while Orwell's nightmare vision anticipated the proscribing of the inner self, the First World was culturally fragmenting, sufficient to give the individual self a fighting chance of visibly asserting itself. No crucible served that function more aptly than Hollywood, where the search for screen presence brought many of its idols under the influence of these neuro-cognitive effects.

According to some sources, 1950's movie star James Dean (1931-55) anticipated the civil rights movement, black power and women's liberation with his vivid depiction of the struggle for personal identity, individuality and self. The 1955 motion picture Rebel without a Cause features James Dean as an emotionally-confused and troubled teenager, whose autonomous morality conflicts with the prevailing social mores – but who nonetheless reconciles himself with his parents, albeit as a consequence of tragedy. Rebel without a Cause paints an authentic picture of the true nature of power, the influence of authority, and an autonomous morality shaped discontinuously by experience. Dean, whose birth was contemporaneous with the invention of the electric guitar, continues to enrich our perception of youth culture, and the complex relationship which exists between young people, power and the law.

Dean's character, Jim Stark, eventually arrives at an autonomous morality which won't simply go extinct – unlike a conditioned conscience, which is little more than a trained reflex. In that respect, there's scope for a youthful testing of the limits within future societies, allied to the sparing use of punishment, especially in those instances where the person's own plasticity provides for an appropriate re-adjustment. This highly-discernable generation gap arose whilst western society was very much under the influence of remote interference, making the two arguably inter-related. But the powers presiding over youth have conflicting priorities, some of which relate to the continuation of governance from as close to the political centre as possible, both domestically and further afield.

Unlike communist states and developing countries, capitalist societies have inherently complex personalities, reflecting their pluralist nature – but when the stakes are very high one senses that terms like determinism, predestination and learned helplessness haven't entirely lost their currency, even in the West. It just so happens that the insipient collapse of the Soviet Union, from the late 1970s onwards, left the young vulnerable to debilitating levels of interference. And, in fact, youth remains vulnerable to the hidden agendas of more powerful others. But, as NATO grows, and the free world becomes increasingly indomitable, the need to reduce, manipulate and control in this way evaporates. Making the future humanizing, rather than dehumanizing.

• **Higher, faster and further**

Rather than pursuing classically-conditioned perfection in the mainstream, decision makers should concentrate on the legitimate use of power, sufficient to justify their standing. In the mainstream this means holding a mirror to those governing, rather than the governed. That wider separation of power within civilized societies ultimately reflects endemic weaknesses across all age groups, and those with adequate plasticity would obviously gain from some form of discontinuous development, whatever their standing. By the 1960s the phenomenon was habitually exaggerating and augmenting individual chemistry – in a way which would land some, not all of them criminals, in prison; and others, not all of them saints, on the surface of the moon. First World history, in spite of, or perhaps because of these controversies, has since taken us higher, faster and further than all the others, whilst leaving in its wake a far more interesting past.

In this context, the right kind of history equates with authentic social progress; with the 1960s being *the* decade, as regards science fully lifting the bonnet on the deepest processes driving society in the modern age. But, due to that process of augmenting, exaggerating and amplifying society's deepest and most pernicious stresses the phenomenon would, paradoxically, force itself to adopt an ever more supranational stance; liberating itself from the domestic tensions of the country which had originally spawned it. But, having elevated itself above national and subnational tensions, its subsequent involvement in the same would appear ever more mysterious due to its self-limiting nature. Mainstream politicians like JFK sought, of course, to hold the balance; often through sheer weight of personality.

It was against this backdrop that the subnational misfit, Lee Harvey Oswald, assassinated the nationally-elected president John F Kennedy; within a mainstream political environment presided over by a supranational presence. A supranational presence which had been present at the untimely death of the international film legend Marilyn Monroe; whose entire **self-concept** had been copiously enriched both nationally and subnationally. That process of enrichment, coupled with the plasticity of their own respective natures, rendered Monroe grossly internalizing and Oswald grossly externalizing. Not all of which was remotely-induced, as evidenced by their gender-specific psychology. Within these two episodes lies an important lesson for us all, in-as-much as multilateral telepathic involvement will actively enrich, sometimes destructively.

It's fair to say that the personal administration of such fundamental aspects of the self isn't easy, especially when faced with the unruly use of power. That's why getting the balance right, as regards the application of such extreme powers, is so important. As for real and growing tensions arising authentically from mainstream differences, these are best resolved at the national or subnational levels, as conflict resolution at a supranational scale may incur a significant loss of life – or simply the loss of a life which is significant. Precisely how many lives, significant or otherwise, were sacrificed in pursuit of substantive political and military solutions, we can only speculate. Seen in this light, the summer of love, flower power and widespread marijuana use, all seem today like a short-lived nervous over-reaction to the same.

As if in deference to the **normal curve of distribution**, America's space program was undoubtedly littered with those who successfully maintained an even-strength – and in spite of events which left many questioning their sanity. There are, of course, implications to maintaining a normal curve of distribution as regards human behaviour, especially when to exhibit the wrong stuff takes as much third party involvement as is needed to demonstrate the right stuff. And, where the wrong stuff is sought only for the purposes of statistical neatness. Extrapolate, and one might expect to encounter a certain amount of rogue behaviour globally. However, without the capacity to engage in punishing neuro-cognitive ill-effects it's unlikely that such behaviour would be anything other than absolutely genuine.

- ### Hearts of darkness

In 1969, American Charles Manson (1934-2017) led four followers into Beverley Hills, where together they brutally murdered eight people over two nights. Multilateral telepathic involvement by both known and unknown third-parties would have been possible. The question is, was there such involvement? One of the reasons this is such a difficult question to answer is that there *are* sociopathic individuals who affront the most elementary forms of morality of their own free-will, leading to general repugnance and horror, and who therefore require little or no encouragement. The danger, however, is that a curiosity-driven, male-dominated phenomenon, hooked on casual involvement, finds the prospect of manipulating vulnerable individuals simply too great. Radiofrequency technologies which have, until comparatively recently, been the preserve of an elite, may have found their way into the hands of more gung ho characters, with more questionable motives. Societies with inadequate gun-control stand to lose most from this unruly arrangement.

In many ways the Manson killings are unusual: today the murderer is more likely to act alone, be heavily-armed, and have no charisma to speak of. A brief appraisal of the available evidence – as regards mass killings, spree killings, running amok, etc – suggests that such events have become more common since the end of the Cold War; with the victims simply blown-away. This transition to cold-blooded murder appears to be contemporaneous with the phenomenon exhausting its original remit. Clearly, individuals should only use firearms to define civilized limits, rewarding the target with their life wherever possible. This argument suggests that only law-enforcement officers and the military require guns, not the general public. Within the NATO community the received wisdom is that gun-control is an absolute must – a conviction which should be allowed to proliferate.

Enriching juvenile thought processes with shallow forms of violence has the effect of making human life seem cheap – to borrow an expression from the celebrated war reporter Morley Safer (who worked for CBS News at the time of the conflict in Vietnam). Those who make human life seem cheap risk becoming architects of their own misfortune. And, if it transpires that our movie-going habits are indeed driven by some heavily-repressed death instinct, then duel inheritance theorists might warn of the dangers of coevolution, i.e. the subtle interplay of cultural and genetic processes – arguing that we're storing-up problems by accommodating rather than correcting such cravings. The ineluctable conclusion is that the marriage of individual minds, without any thought as to culture, is probably unwise.

•   Soup cans amid fragmenting subcultures

The summary execution of Marxist revolutionary leader Che Guevara (1928-67) at the hands of Bolivian soldiers neatly divides modern from postmodern: his posthumous image borrowing heavily from low-art and popular culture (in the manner of **Pop Art**) rather than from the previous fine art traditions. Western culture was already fragmenting into a multitude of genres and subcultures prior to the postmodernist era, but that process began to accelerate rapidly from the mid-1960s. These microsociological sub-cultures often acting as a counterpoint to much blander and more hackneyed macrosociological trends and influences. Unfortunately, neither individual subcultures nor society as a whole had a perfect handle on the deeper realities permeating western civilization at that time, so that one might find equal truth and meaning at all levels.

The law uses terms like realism, positivism and natural, and in many ways we can appropriate these expressions for the purposes of examining culture. For example, **natural culture**, or culture at its most natural, is practical and commonplace (e.g. flushing toilets, brickwork and the internet); **cultural positivists**, on the other hand, would perceive culture as that which is posited, but not necessarily experienced (e.g. immersive virtual reality art); whilst **cultural realists** would say that culture is what one actually meets with, or gets, in the course of going about one's business (be it architecture, religious idolatry, or simply chewing gum on the pavements). What appears to be happening is that a lot of things are being posited as being our culture, which we don't commonly meet with, or even know exist. More worrying, coevolutionists might warn that the present culture of chewing gum on the pavements is feeding back into the gene pool.

Artists like Andy Warhol (1928-87) understood this implicitly; casting his Pop Art in the mould of culture at it most natural or commonplace (e.g. the ubiquitous soup can), precisely because most art is invisible once we get beyond the neatly stacked rows of Campbell's soup. Also, many artists in the 1960s may have begun to feel alienated from deeper truths, rather than in touch with the same – making the superficial a more pertinent point of reference. If so, conceptual art, and the so-called installations which followed, may be seen as a response to that alienation. We can only speculate as to what artistic traditions will follow in the future, if deeper truths do ultimately reveal themselves.

One would expect the effects of a paradigm shift to be reflected in the work of major artists. At present one is more likely to see evidence of the need for the same in the artist's general disposition, mood-swings or apparent drunkenness, rather than in their installations, compositions or conceptual arrangements. Popular music has been the most courageous art form of them all – broaching these issues more than any other artistic medium, through its use of metaphor and allusion. So much so, that while conceptual art often finds itself dismissed as superficial nonsense, posterity will appraise popular music far more sympathetically. Having said that, the abstract artist may argue that their art is a true reflection of their own inner self, whereas the musician's lyrical symbolism derives from exogenous factors impacting on the same. This permeation of induced effects into popular culture is a reflection of their application from a relatively young age. A significant percentage of a given cohort of school leavers will be profoundly affected, but not always visibly. A minute fraction of those will pen lyrics with a self-conscious intensity – the rest will run the gauntlet of such intriguing, becoming, amongst other things, drug-addicts, prostitutes, and so many disaffected males. If extrasensory effects are assimilated into mainstream society, then they will become a natural part of human existence, like air travel, secondary education and dual carriageways. If they are posited, but not experienced, some will argue that they aren't even part of mainstream culture at all. And, viewed realistically, if they are experienced in an unruly manner, that will serve to create the most unenviable of cultures.

Sharing a common experience has bound people ever since prehistory, when Stonehenge became seasonality set in stone; effectively breaking down the mystique surrounding nature in a mutually stimulating manner. By breaking down the barriers between fine art and low art, postmodernists provided for the establishment of the present **Turner Prize**, and with it the likes of Damian Hurst (1965-), Grayson Perry (1960-), and Tracey Emin (1963-). In other words, the kind of people who might gain inspiration from multilateral telepathic involvement, hypothalamic confusion and the gross internalization of their issues. In fact, the Turner Prize has become so prestigious that it's the final word as regards the positing of culture – so much so, that one can't realistically hope to avoid it.

- ### The neuro-cognitive lingua franca

The significance of English worldwide owes much to the British Empire in the 19th century and US power and influence in the 20th century. For its part, the western media has finally abandoned its bias towards **received pronunciation**, increasingly allowing regional accents and vernacular phraseology to flourish. All of which mirrors the gradual break-down in the class system and the emergence of a generation gap. That's not to imply that the West has altogether abandoned its use of **prescriptive rules** (i.e. recommended practices) and **proscriptive rules** (i.e. things to avoid), it's simply that it's now much more sparing in its criticisms. Therefore, English looks set to become the official language of telepathic communication – making reasoning with the devil that much easier, should we carelessly create one.

The news media provides a convenient benchmark as regards written and spoken English. In that respect, newspapers and news reporting serve us in multiple ways, one of which is presenting the language in its most received form. As such, it is arguably the best evidence there is of grammatical change, vowel shifts and contemporary style. Several large news agencies, e.g. **Associated Press**, based in New York, and **Reuters**, based in London, act as primary sources of news. These global agencies supply news from around the world to thousands of newspapers and hundreds of radio and TV stations. In this way, a broad historical narrative is slowly built-up.

Various theories attempt to explain mass communication, some of which have implications for those engaged telepathically. For example, **limited effects theorists** stress the importance of personal preference, as regards how we access and perceive information, which greatly limits the impact of the same. Other theories include **selective retention**, whereby people are said to remember the information which is most consistent with their existing attitude. **Reinforcement theory** suggests that one's perceptions are simply reinforced. And, **two-step flow theory** implies that intermediaries actively mediate the message coming from the primary source. All these have something to say about multilateral telepathic involvement, and the way in which people build-up mental images using extrasensory information.

The question arises as to whether the government should be an instrument of the people operating in a cold-blooded computational fashion, or whether the people should be an instrument of the government, given their deference to its privileged knowledge and processing power. In reality, this particular model cannot hope to mirror the all too human conjuring with information which actually goes on in government; and one must conclude that man and machine are so dissimilar that those differences are likely to be amplified, not weakened – especially in democratic states, whose complex personalities beg profound attributional questions.

•   Supranatural selection

A process of **supranatural selection** now dominates evolution in the **anthropocine**, resulting in the survival of the most well-balanced. This is particularly important as we approach a post-peak future filled with energy insecurity and resource scarcity. And, it more or less guarantees that apathy on issues as serious as global warming isn't simply a choice, but a serious disadvantage. Looking further ahead, into the more distant future, the whole of the free world could be systematically re-configured, if necessary, enabling it to absorb the more ruinous events encountered in cosmology, climatology and volcanology. Destructive interference on such an enormous scale would be a formidable test of humankind's ability to manage both itself and its environment. Such anticipation and preparation runs contrary to our instincts, but balanced-mindedness nonetheless demands it.

The 20th century was principally **reductionist**, as scientists sought to understand such things as light, atoms and quanta; but the coming century is likely to be far more **holistic**, as experts in various fields pool their collective knowledge and integrate their various branches. Significantly, human psychology continues to defy full scientific explanation; psychology often being perceived as a pseudo-science, rather than an orthodox science. The explanation presented within these pages is that human psychology is inherently elusive, due to the elasticity afforded by its various parts. And that if one replicates that psychology at the level of states, with multilateral telepathic involvement mirroring the absorption and emission of signals within a given mind, then the personality of the nation becomes equally intangible. Some might term it chaotic; implying that the future is unremitting chaos, alleviated only by the constitutional structure.

That intangibility does at least keep mankind on its toes, driving ingenuity and invention through the infinite scenarios it then conjures-up. In the final analysis, we can no more predict the behaviour of a truly democratic nation, than the actions of a given individual. By imbuing the state with the characteristics of the human psyche we find ourselves deeply embedded within an organ possessing as much flexibility as its constituent parts; making that plasticity work for us, not against us. With this in mind, this well-intentioned polemic ends with a brief examination of life, the universe and everything. Including, how everything could have sprung from nothing and how telepathy is evidence of an **anthropic principle** pervading the entire cosmos. Plus, we see how the forces driving cosmic inflation lead naturally to a self-limiting 'creator', acting mysteriously in the wider interest, via ubiquitous DNA.

Ch18: Equilibrium

- DNA

Deoxyribonucleic acid (DNA) is person-specific, just like the brains that it fabricates. DNA is therefore doing its bit to fill-in the gaps in the electromagnetic spectrum, with consciousness being a miraculous side-effect of that ongoing astronomical process. Atoms of the elements carbon (C), oxygen (O), hydrogen (H), phosphate (P), and nitrogen (N) are all that one needs to construct this molecular blueprint, from which all known life is fashioned. Nucleic acids are **polymers**, comprising large molecules made up of chemically-bonded smaller molecules. In DNA these smaller molecules are termed **nucleotides**. So, a whole string of nucleotides, covalently bonded together, creates a nucleic acid. As for deoxyribose, that's just one of several parts making-up each nucleotide.

In fact, nucleotides have three essential components: 1) a **nitrogenous base** (which can be either guanine, cytosine, adenine, or thymine); 2) a **sugar** (deoxyribose); and, 3) a **phosphate group**. Thus, there are four distinct nucleotides, whose identities are defined by their nitrogenous bases. These individual nucleotides are joined together via **phosphodiester bonds** – which is just a technical way of saying that a phosphate group on one nucleotide, covalently bonds to the sugar of an adjacent nucleotide. In this way, two nucleotides are joined together. Now continue to add a further 130 million or so nucleotides, in order to create a single strand of DNA. Just remember, you'll need to covalently bond each phosphate group to the sugar of its neighbour.

This sequence, comprising a single linear strand of DNA, will ultimately determine the characteristics of the life-form it gives rise to, be it fauna or flora. The helical DNA we hear so much about comprises two of these strands. Each strand comprises the same basic building blocks, and complements the other, in-so-much as **base-pairing** demands that only two base-pairings are possible between nucleotides on opposite strands – guanine (G) with cytosine (C); and adenine (A) with thymine (T). These two strands are joined at these base-pairings by H-bonding. Typically, the energy associated with covalent bonding is more than ten times greater than H-bonding. Therefore, each individual strand is significantly stronger than the forces holding the two together.

Structurally, each strand is a distinct macromolecule, more weakly bound to its twin via intermolecular forces. Bound together in this manner, the secondary structure adopts the iconic double helix shape (see Fig. 32), not least of which because the base pairs are hydrophobic (and are shielded from the aqueous environment of the cell), whereas the hydrophilic sugar and phosphate groups (making-up the backbone) are exposed to it. DNA has been copying itself ever since life evolved – through a process known as **semiconservative replication**; so-called, because the DNA double-helix 'unzips' and each strand begets another DNA helix through the attachment of free nucleotides. For this to happen, the cell requires: 1) intact DNA; 2) free nucleotides; 3) the requisite enzyme (DNA polymerase); 4) energy (ATP); and, 5) a moderate temperature.

Figure 32: DNA (double helix)

Phosphodiester bonds create two linear strands, which are joined together via nucleotide base-pairing (see text).

DNA is the blueprint for all organic life – not simply when reproducing itself during sexual reproduction, but also throughout the whole process of development, from zygote (egg) to mature adult. A person's DNA is sited in their **chromosomes**. The chromosomes being rod-shaped structures containing the entire genetic code, or **genes**, for a given individual. A normal human cell contains 46 chromosomes arranged in pairs (hence there are 23 *pairs* of chromosomes). One can define a gene as being a sequence of **bases** needed to produce a specific physical characteristic or physiological function. Consequently, there are multiple genes on a single strand of DNA (and therefore the chromosomes contain all the genes for a given individual).

Sexual reproduction brings together the **gametes** from both the man and woman (spermatozoon and ovum, in the case of humans). Unlike all other cells in the human body, these gametes have only 23 chromosomes apiece (making them haploid in number). However, when they fuse into a zygote, the resulting egg contains 23 *pairs* of chromosomes (in other words, the requisite diploid number). This fertilized egg (or ovum) then grows into a person possessing a complex anatomy, e.g. skeleton, muscles, central nervous system, brain, eyes, etc. As all these various bodily structures contain precisely the same genes, a process of **differentiation** is called for, allowing otherwise like-cells to differ. This process of differentiation is poorly understood. What we do know is that processes in the cell's nucleus or cytoplasm switch genes on and off, giving rise to a whole range of bodily structures – including sexual characteristics, such as those appearing in puberty.

Of the 23 *pairs* of chromosomes commonly found throughout the human body – one pair are the **sex chromosomes**; whilst the other twenty-two pairs are common to both sexes. These sex chromosomes determine the sex of the individual. In humans, the sex chromosomes are labelled X and Y (in other words there are two types). Women possess two X chromosomes, whereas men have an X and a Y chromosome. Gamete formation in men produces sperm cells with only 23 chromosomes – rather than the full complement of 46 (arranged in pairs). Half the spermatozoa will therefore possess the X chromosome and half the Y. If a sperm cell carrying a X chromosome fuses with an ovum (which can only possess an X chromosome) then the resultant child will be a girl (X-X). If the sperm cell is carrying a Y chromosome, the result will be a boy (X-Y).

Thus the man determines the sex of the child through his sperm. Whilst the whole sequence of nucleotides and the genes they contain determines the resultant child's general morphology, plasticity, and to some extent their psychology. Ultimately, we can either manage the gene pool or manage whatever the gene pool naturally gives rise to. The model presented in this book suggests that much can be done to manage whatever that net biological inheritance happens to be, through the concentration, separation and denial of power. One thing is for certain, the aforementioned supranatural presence is a far stronger evolutionary force than mainstream popular culture.

The state, for example, is a person in the eyes of those applying the effects, leading to the evolution of individual geopolitics. This is evolution not by **neo-Darwinian synthesis**, but by supranatural selection. Such selectiveness has favoured the Allies over the Axis Powers, the Western Allies over the Soviets, and due process over the alternatives. Human DNA will one day reflect these choices; including our commitment to telepathy, legitimacy and balanced-mindedness. But in spite of history's endless tides, people will always be at liberty to think with impunity – now, tomorrow and always.

• Biological inheritance

The biologist Lewis Thomas (1913-93) once wrote: "The capacity to blunder slightly is the real marvel of DNA. Without this special attribute, we would still be anaerobic bacteria and there would be no music". It would be strange indeed if all that semiconservative zipping and unzipping didn't lead to occasional blunders – due to irregular conditions, non-availability of free nucleotides, or simply the destructive effects of ionizing radiation. These blunders amount to **genetic mutations**, i.e. slight changes to the genetic sequence, base-pairings, or a permanent alteration to the way in which genes are switched on or off (causing the organism to grow in different ways). One might argue that the mechanism behind evolution is thus evolving, as the range of factors imposing on DNA rises through time. A successful mutation will find replicating itself easier; but what is successful remains somewhat relative.

When we think of the theory of evolution we think of the British naturalist Charles Darwin (1809-82). In Darwin's view, organisms produce more offspring than the environment can support and those with favourable characteristics survive – resulting in the evolution of new species, perfectly adapted to their environments. Darwin understood implicitly, as any ecologist will attest, that all species will eventually meet with **environmental resistance**, be it climate change, food shortage, or predation (due to continental drift, volcanism, competition, or, indeed, any number of exogenous factors). Evolution is therefore the only practical mechanism which life has to surmount these obstacles. Thus evolution represents a continuous re-adjustment, by living-organisms, to various forms of environmental stress.

Human prehistory witnessed the exploitation of the planet's available space, just as history has seen organized societies dominate a given space through time. Finite space and unlimited time now define the personality of the global civilization we presently synthesize. Any form of resistance which we now encounter, which is so great it cannot be overcome within the parameters of the present genome, will force mankind to evolve. Multilateral telepathic involvement may serve to create such a stress. And, it's fair to say, this could significantly impact on the **human genome** in the long term. But a distinction needs to be made between the discomfiture of multilateral telepathic involvement, and those blatantly punishing ill-effects which one proposes to prohibit.

With the phenomenon suitably revised, the neuro-cognitive biosphere should be inherently well-balanced – not something the genome in its present form should find too onerous. Seen in this light, the phenomenon's replacement, NEURON, might seem like an ally, helping to craft a superior mainstream global environment; and assisting mankind to surmount environmental stress without recourse to actual evolution. Arriving at this position has involved deception, sufficient to bring about disequilibrium within the sciences. Disequilibrium being synonymous with growing incommensurability and the need for a new paradigm. To restore equilibrium we need to re-evaluate the truth, free of the demands of 20th century political and military intrigue.

For this reason, the remainder of this chapter delves into some of the deepest and most intriguing mysteries surrounding time, space and matter – in order to see whether we can shed any new light on the same, in the light of what's been said. One thing we can say, with absolute certainty, is that all that is spoken of throughout this book is impossible without the prerequisites of **space** and **time**.

•   Space and time

Ontology asks: what actually exists? The presumption must be that everything that exists requires space. Particles are points in space, and we know they must be moving, because if they weren't they'd require no space at all. Atoms and molecules arrange themselves across space, and their interactions are mediated by a range of forces. Forces which either 'bite' and effect a change (i.e. **coupling**), or which fail to bite, or stop biting (i.e. **decoupling**). That specific biting point (the **coupling constant**) being the dividing line between the two alternate states. According to orthodox science, the principal forces in nature are the electric, magnetic and gravitational forces – together with the strong and weak nuclear forces. Both gravitational forces and electromagnetism operate over astronomical distances; whereas the strong and weak nuclear forces are said to operate over very short distances, at the sub-atomic level.

Coupling gives rise to atoms, molecules and people due to the combined strength of all the relevant forces. For example, when the strong nuclear force bites it concentrates points of positive charge in the nucleus of an atom – namely, protons (together with a given number of neutrons). When this strong nuclear force stops biting, as happens in a nuclear explosion, the concept of **equivalence** takes over, and one is faced with so much blinding energy. That nuclear explosion amounts to a very sudden and highly-dramatic decoupling of the strong nuclear force; the impact of a single neutron having caused the strong nuclear force to dip below the coupling constant for that particular nucleus. This creates space between the points of positive charge.

One requires only a child's U-shaped magnet half-dipped in orange paint, and a paperclip, to understand the mechanism behind **space-creation**. The paperclip is either picked-up by the magnet, or it isn't; with the amount of magnetic force needed to bring the two together being the coupling constant. To create space, simply pull the two apart. Thus, space is simply what happens when the principal forces in nature stop biting – and time is simply a side-effect of distance. Clearly, for time to exist, one needs space; for nothing can exist without it, not even time. Thus, space and time are aspects of a mutable phenomenon (one which has switched from one state to another).

Prior to decoupling, there was only coupling. Neither time nor space existed; and the point at which all the forces were biting, conventionally termed a **singularity**, couldn't move, because there was no space. Technically-speaking, singularities don't exist, because everything which exists requires space! Eventually, that state of relative non-existence collapsed through decoupling, whereupon space came into being and universal inflation took off – and with it the space, time and materials needed to create atomic nuclei and their orbiting electrons.

According to orthodox science, anything – you and me, for example – arise because nature's principal forces are trying to create nothing in empty space. A mathematical proof might one day be arrived at, akin to **Fermat's last theorem**, which proves that something can, in fact, spring from nothing. And that you, and all the people mentioned in this book, have, by your very existence, frustrated mother nature's attempts at stamping-out time, removing space and fabricating nothing whatsoever. At the heart of this cosmological paradox lies the expansion of the universe, whose underlying mechanism forces things to exist.

- ## Coulomb's Solution and the Dark Cycle

French physicist Charles de Coulomb (1736-1806) arrived at a law, which states that the electric force of attraction or repulsion is proportional to the product of the charges, and inversely proportional to the square of the distance between them. Additionally, his law for magnetism states that the force between two magnetic poles is proportional to the product of their strengths, and inversely proportional to the square of the distance between them. Both of these laws bear a striking similarity to Sir Isaac Newton's law of gravitation, which states that the force of gravitational attraction between two objects is proportional to the product of their masses, and inversely proportional to the square of the distance between them. Together, these three equations suggest an answer, as regards the mechanism driving **cosmic inflation** (see Fig. 33).

Figure 33: Dark matter (missing mass) - Coulomb's solution

Force (F)	Equation	Coupling constant (the either/or result or action)
Gravity (Gravitation)	$F = G \dfrac{m^1 m^2}{d^2}$	The photon (emitted by m^2) is either: (a) absorbed by m^1, or (b) is inhibited in its line of travel; i.e. not absorbed by m^1
Electric (+/-)	$F = C \dfrac{q^1 q^2}{d^2}$	The ion (q^2) is either: (a) electro-statically bound to q^1, or (b) remains free of q^1
Magnetic (N/S)	$F = K \dfrac{M^1 M^2}{d^2}$	The paperclip (M^2) is either: (a) picked up by the U-shaped magnet (M^1), or is (b) not picked up

In each of the three examples, quoted in the table, there are two alternate states, corresponding with coupling and decoupling. These two distinct states are separated by a coupling constant.[60] Where the magnitude of the force (denoted by the symbol F) is less than the relevant coupling constant, light cannot flow between the two objects (m^1 and m^2); the ions (q^1 and q^2) remain un-attached; and the paperclip (M^2), lying on a surface in the presence of a U-shaped magnet (M^1), remains immobile on that surface. So, relatively weak electric and magnetic forces give rise to so much space between objects – and, whilst this isn't entirely the mechanism by which the fabric of space is created, it is nonetheless related to that particular physics.

Actual space, as in the invisible medium which delineates the limits of the cosmos, and which permeates every single atom, requires both electromagnetism and gravitation for its production – **gravitation** being weaker than gravity (the two being separated by the coupling constant for the gravitational force). Whilst gravity supports the free movement of photons, and therefore enables objects to be seen, gravitation actively inhibits the same. This means that the universe is an assortment of both visible and invisible objects. Wave-particle duality enables us to think of light as a photon, wave, or even oscillating electric and magnetic fields. Whichever definition one chooses, gravitation suppresses the free movement of radiant energy and triggers its conversion into the very fabric of space (see Fig. 34). One has termed this process, the **dark cycle**.[61]

Figure 34: **The Dark Cycle**: Runaway Inflation

Where the conditions for the continued movement of a photon cease, the photon ceases being a quantum of action and becomes, instead, a quantum of inflation (see below)

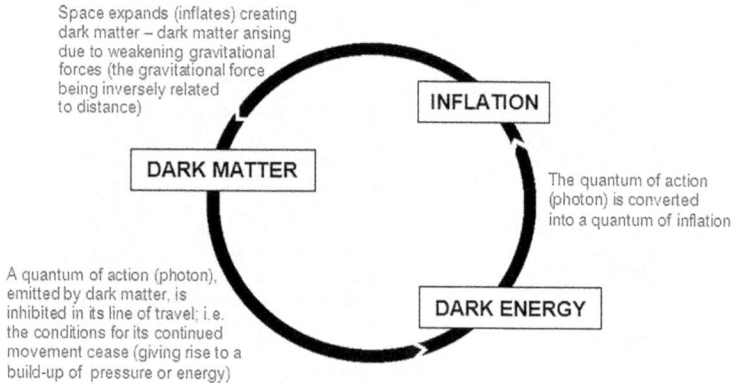

Space expands (inflates) creating dark matter – dark matter arising due to weakening gravitational forces (the gravitational force being inversely related to distance)

DARK MATTER

A quantum of action (photon), emitted by dark matter, is inhibited in its line of travel; i.e. the conditions for its continued movement cease (giving rise to a build-up of pressure or energy)

INFLATION

The quantum of action (photon) is converted into a quantum of inflation

DARK ENERGY

The dark cycle implies that the photon is a point in space, produced when the electric, magnetic and gravitational forces bite, and which moves through space independent of the light source. The wholesale transduction of this **dark energy** (and associated radiation pressure) into the very fabric of space forces **galaxy clusters** apart and renders **dark matter** objects invisible. Because the gravitational force is inversely-related to distance, the expansion of the universe forces the conversion of yet more energy, giving rise to a runaway effect. All the energy directed at us by dark matter or missing mass is converted in this way, i.e. into inflationary effects – leaving none remaining to be picked-up by earth-based telescopes and dishes.

On the face of it, the universe is heading for peak gravity, after which gravitation will begin to dominate. That moment may already be upon us, as expansion seems to be accelerating, arguably due to the conversion of increased amounts of radiation. As luck would have it radiation is electrically neutral, and its transduction into space provides the perfect medium for electrical phenomena to be supported, free of superfluous noise. Within this electrically neutral milieu points of positive and negative charge combine in strict proportions to produce electrically-neutral atoms. These atoms, and the molecules they produce, absorb and emit radiation, much of which expands the cosmos. Once lost to space, however, this energy has the appearance of not being recovered.

•   Action at a distance

A photon travels independent of its source across space and time, with the precise amount of energy lost to space when the photon disintegrates being given by the equation $E = hv$ (where v is the frequency and h is **Planck's constant**). Whether all that lost energy is equivalent to all that gravitational attraction across space and time is an open question. The principal particles possess sufficient mass to be affected and influenced by gravity (and, in the form of atoms and molecules, add to that medium through emission). This synopsis hints at electromagnetism being the most primary of all the forces; with all that exists deriving from that original burst of energy.

The gravity which binds the visible universe derives from the fabric of space. That universal fabric being an isotropic medium which is uniform in all directions, electrically-neutral, and supportive of the propagation of light-waves in a regular and predictable manner. Thus space-creation forces objects apart, provided those gravitational forces aren't substantial. Conceivably, this so-called Higgs field might imbue particles with mass, concentrating the same through fusion; and all the while supporting the electric and magnetic fields (and, indeed, the whole of electromagnetism). It is these, and closely-related fields, which determine how the electrons arrange themselves within space, and which provide for chemical bonding.

Provided the resultant atoms and molecules remain bound by gravity they can share emitted energy; giving rise to numerous open, closed and semi-isolated systems. Increasingly, these open, closed and semi-isolated systems have become estranged by gravitation, giving rise to isolated systems, conventionally termed

227

dark matter. However, within the visible universe, bound by gravity, semi-isolated systems give rise to complex neurology. Intelligent life, like ours, which is able to discern a universe inflating under the influence of its own radiation and which seeks to reflect, as far as possible, a model of balanced-mindedness provided for by the most elementary of cosmic forces.

• **The last one hundred years**

The cosmos is extremely good at creating the chemical elements needed to produce life, but not very good at destroying the chemical basis for life. What's happened is that the whole sequence of star formation and stellar death has thrown out a rich source of chemical elements, into an environment which is incredibly stable in terms of its physical reactions. Even though **chaos theory** proposes that a few simple laws can result, over time, in unfathomable complexity; those initial laws, at least, remain really rather simple. So that in spite of the fundamental realities of nature seeming complex and unfathomable, we nonetheless have the means to impose upon them profoundly, at the level of these basic laws; thereby adding to that immeasurable confusion (such are the blinding ramifications of ecological and climatological disruption).

Remote manipulation, mental telepathy and the superimposition of persons are all products of base-chemistry, but the ramifications associated with their use may transcend human intelligence – unless we frame their use wisely. The last one hundred years has witnessed the global concentration of such powers, and in a manner which could serve humanity well. With this in mind, this book has attempted to fairly appraise the recent past, not least of which, science itself. It has examined what science has done, how science has advanced, and, perhaps most significant of all, what science ought to be doing. It remains to be seen, however, whether the coming century judiciously builds on all that has been learned and experienced. Well-intentioned deception remains possible; and the real genius of science, and certain scientists, has been that they haven't always been forthcoming with the truth.

The 20th century was a battle fought with ourselves over the direction of the human spirit. Now we must enrich our understanding of the past, in order to make the process of remembrance all the more poignant. NATO is the key to strategic stability in the future, its growing sphere of influence being the free-world. The whole balance of power has now shifted in the free-world's favour, making now the right time to engage people openly on these issues.

One hundred years from now people will be facing a post-peak future, and it's important they inherit advantages, derived from our actions, just as we have benefited from the actions of those who've preceded us. Having said that, and leaving aside the caveat that both science and the universe are deceptive, human advance is really quite something, given that reality constitutes a whole series of forces trying desperately to produce nothing, and often in ways which appear prima facie destructive.

Appendix

Glossary

All listed in alphabetical order, ignoring the prefix 'The' (e.g. The Hague appears as 'Hague, The'); and ignoring any preceding numbers, e.g. 9/11 terrorist attacks is listed as 'Terrorist attacks, 9/11'.

A	Definition / description
Absorption profile	The person-specific pattern of absorption and emission which arises due to the influence of personal biomagnetism. Such a profile provides for radiofrequency technologies accessing the mind's of individuals via electromagnetic waves.
Abu Ghraib Prison (Iraq)	Renowned journalist Robert Fisk (1946-) reported on "Saddam's gang-rape and torture in Iraqi prisons" in the mid-1980s. Two decades later there were reports of similar abuses in Abu Ghraib Prison; this time committed by rogue elements within the US military – such abuses making the military seem dissolute, rather than heroic.
ACP	The African, Caribbean and Pacific Group of States (ACP) comprises 48 sub-Saharan nations, 16 Caribbean nations, and 15 Pacific nations. All these countries are bound to the European Union through the Cotonou Agreement. The alliance seeks to reduce inequality, facilitate global economic integration, and encourage sustainable development.
Action potential	Stimulation results in the movement of positively-charged sodium ions into a given nerve cell. This sudden depolarization of the neuron's membrane may exceed the threshold potential for the said cell, triggering a nerve impulse (otherwise known as the action potential).
Activation energies	The amount of energy which must be present in order for molecules to take part in a chemical reaction. The activation energy strongly correlates with the speed of the reaction – the lower the activation energy, the faster the reaction.
Actor-observer bias	Occurs when negative behaviour in others is attributed to their character, but the same behaviour arising in oneself would be attributed to the demands of the situation.
***actus reus* (Latin: the guilty act)**	An essential feature of any crime is the act itself. Actus reus includes commission, omission and possession, i.e. committing something, omitting to do something, or possessing something.
ADHD (Attention Deficit Hyperactivity Disorder)	The most commonly diagnosed childhood disorder in the USA; ADHD is characterized by an inability to follow instructions, complete tasks, pay attention and sustain conversations. As with all disorders its imperative that they are authentic and innate, particularly as children are inherently vulnerable to radiofrequency effects.
Adsorption	The mechanism by which two reactants are temporarily bound to the surface of an enzyme, thereby distorting and weakening their covalent bonds, sufficient to speed-up the reaction time.
Afferent signals	Nerve impulses traveling towards the brain.
Agenda-setting	Determining what the priorities are in term of topical debate and mainstream decision making.
Agent provocateur	A person or body which actively encourages a crime to be committed, with the intention of securing a conviction or effecting some action (possibly military). This includes placing a person under such severe stress that a breach of the law is likely to result.
Alpha particles	Two neutrons and two protons, i.e. a positively-charged helium nucleus. Many radioactive substances emit alpha particles.

al-Qaeda	The anti-Western Jihadist movement which grew out of the conflict in Afghanistan, during the Soviet occupation of the country (1979-89). Comprising so many disaffected Arab males, and heavily financed by Saudi money, al-Qaeda became the pre-eminent terrorist organization of the 1990s. The organization responsible for the 9/11 attacks.
American Civil War (1861-65)	The issue of slavery and federal union tore the Democratic Republican Party apart in 1854; and it wasn't long before America itself was torn apart by these very same issues. The American Civil War lasted from 1861 to 1865, and served to define the future character of the USA, i.e. strong federal government and the codification of equality.
American Revolution (1765-89)	The American Revolution began in 1765 amid American colonist's complaints that direct taxation from Britain was a violation of their rights. By 1775 the colonist's were mobilizing a militia amid British troop reinforcements. Much of the resulting American Revolutionary War (1775-83) was fought in the wake of the colonist's Declaration of Independence (1776). This war proved un-winnable for the British, who capitulated with the Treaty of Paris in 1783. The revolution ended in 1789 with the inauguration of America's first president, George Washington, and with the US Constitution coming into force.
American Revolutionary War (1775-83)	War fought between the 13 English-speaking colonies on America's eastern seaboard and the British, due to irreconcilable differences over representation and taxation. See American Revolution.
Amplitude	The maximum displacement of a wave from what would otherwise be a flat calm or rest position.
Amplitude modulation	A means of impressing information onto a carrier wave, by varying the wave's amplitude (or displacement from flat calm).
Anglo-Irish War (1919-21)	The so-called Easter Rising, which took place in Ireland in 1916, was demonstrative of rising tension – tension which spiralled into the Anglo-Irish War. This war ended with the partition of Ireland (in 1921), resulting in a protestant-dominated Northern Ireland in the north and a catholic-dominated 'Irish Free State' or Eire in the south.
Anomaly	Irregularity, peculiarity or inconsistency, i.e. deviating from the norm.
Anorexia nervosa	Condition mainly affecting young women, characterized by self-induced weight loss and distorted self-image. Historical accounts dating back centuries suggests causes other than simply modern living; though modern living and radiofrequency effects may serve to exacerbate any underlying predilection.
Anthropic principle	Expressed in its stronger form the laws of the universe arrange themselves in ways which inevitably give rise to humans (or the intellectual equivalent). Conceivably, nature exploits the gaps which exist in the electromagnetic spectrum creating complex neurology by default. If so, humans arise due to the chemistry and physics filling those gaps – and in a manner which is quite elemental.
Anthropobionomics	The study of Homo sapiens in relation to its environment.

Anthropocine	Present geological age in which mankind's activities serve to affect the nature of the rock strata subsequently laid down. As rock strata, heavily influenced by the human impact, has yet to surface, geologists have plenty of time to debate the actual start date.
Antiparticles	A particle which has the same mass as its counterpart, but differs in every other respect, having, for example, the opposite charge. Although annihilation is said to occur (producing energy), equal and opposite charges commonly coexist; begging the question, to what refutable hypothesis are we to attribute this mutual obliteration.
Anti-Semitism	Actions, policies and practices which discriminate against Jewish people.
Anxiety	Anxiety becomes a problem or a disorder when there is no obvious cause. It is possible to superimpose anxiety symptoms, possibly symptoms arising in one's own person on an earlier occasion (in response to some tangible or overt stress).
Aqueous solutions	Solutions such as those found in living cells, which consist largely of water. Various substances can be dissolved in water through the formation of hydrogen bonds, arising between the substance (solute) and the dipolar water molecules (solvent).
Arab-Israeli conflict	Following Israel's creation in 1948, displaced Palestinian Arabs drew much sympathy and support from neighbouring Arab states, creating a continuous state of political and military tension in the region. This tension has resulted in several localized conflicts (most notably in 1948, 1956, 1967, 1973, 1982, 1986, and 2002). This ongoing conflict has necessitated numerous UN Security Council resolutions.
Aristocracy	The worst accumulations of wealth and inherited privilege in the historical past gave rise to an upper-class, or aristocracy. Such privilege and wealth has largely been confined to history through social and political reforms, but gross inequality remains a concern.
Arrow Cross	The effective annexation of Hungary by the forces of the Third Reich in 1944 emboldened Hungary's very own Arrow Cross militia, whose fascist aims and anti-Semitic ideas complemented those of Nazi Germany.
ASEAN	The Association of Southeast Asian Nations (ASEAN) comprises Indonesia, Malaysia, Philippines, Singapore, Thailand, Brunei, Burma (Myanmar), Cambodia, Laos and Vietnam. Economic partnership, peace, security and progress are central to this alliance.
ASEM	The Asia-Europe Meeting (ASEM) comprises the whole of the European Union, Russia, Mongolia, China (PRC), Japan, South Korea, India, Pakistan, Australia and New Zealand (plus several other countries). It exists to strengthen relations in areas such as politics, economics, culture, society and education.
ASPA	The Summit of South American-Arab Countries (ASPA) comprises the League of Arab States and the countries of South America; which together cooperate in areas as diverse as economic development, science and technology.

Asperger's syndrome	Some texts suggest that Asperger's syndrome is simply an autistic temperament combined with normal or high IQ. Thus socially gauche high-achievers, substituting academic interests for ones demanding empathy and involvement, fit this description.
Assimilation	The process by which new ideas and practices are incorporated into an existing structure, without fundamentally altering that which is doing the assimilating.
Associated Press	Major news agency (based in New York, USA) which acts as a primary source of news material for newspapers, TV companies and radio stations globally.
Atheism	The belief that there is no supreme deity or God.
Atom	Atoms comprise three major particles: protons, neutrons and electrons. As the number of protons and electrons is equal the atom carries no overall charge.
Atomic mass units	The relative atomic mass of an element is expressed in atomic mass units (amu). Atoms of the same element can have different masses, due to differing numbers of neutrons, resulting in a weighted average.
Atomism	Theory that everything is made of atoms, and that things are best explained by reference to them.
Attribution theory	This theory is concerned with how people attribute causes to their own an other people's behaviour. Its importance lies in the fact that attribution is very often fraught with subjective bias, egocentricity and fundamental errors.
Audience effects	The effect of being watched in which some actions and behaviours become repressed, whilst others are perhaps strengthened.
Aufbau principle	The principle that electrons will fill the orbitals in an atom sequentially – starting with the lowest energy orbitals, closest to the nucleus, first.
Autism	Characterized by abnormal social interaction, inability to form friendships and repetitive forms of behaviour. The condition is more common in males, possibly due to the fact that girls have cerebral hemispheres which communicate with each other more effectively.
Autonomic nervous system	That part of the peripheral nervous system which is subdivided into the sympathetic and parasympathetic nerves. These two divisions act together, with the main centres located in the hypothalamus. Together they control blood pressure, body temperature, heart-rate, etc.
Autonomous morality	Morality which derives from one's own character, and which is supplemented discontinuously by experience and learning. Such morality is unlikely to go extinct, unlike heteronomous morality, which is no more than a conditioned response.
Axon	The axon serves as a transmitter in a neuron, conveying nerve impulses away from the cell body to other neurons. Conversely, dendrites receive signals from other neurons.
B	
Balanced-minded	See well-balanced.

Balfour Declaration	The Balfour Declaration (1917) was a letter, by the then Foreign Secretary, favouring the establishment of a national Jewish homeland in Palestine (a mandate subsequently appropriated by Britain at the conclusion of the First World War). This commitment, later codified in the British White Paper of 1922, shocked Arabs who'd risen against the Turks at the behest of T.E. Lawrence, and who expected only self-determination, independence and autonomy.
Balkans	A peninsula in south eastern Europe, with the Adriatic and Ionian Seas to the west and Aegean and Black Seas to the east. Includes many politically volatile regions and much ethnic and religious diversity.
Base-pairing	DNA's nitrogenous bases come in four varieties, i.e. guanine (G), cytosine (C), adenine (A), and thymine (T). Two complimentary strands of DNA are bound together, into a double-helix, via their nitrogenous bases – however, only the following base-pairings are possible, A-T and C-G. A small sequence of these base-pairings, along an entire length of DNA, constitutes a gene.
Bases	See nitrogenous base.
Behaviourism	A branch of psychology which was popular in the first half of the 20th century due to the influence of Russian physiologist Ivan Petrovich Pavlov (1849-1936) and American psychologist John Watson (1878-1958). Both saw overt behaviour as the only useful yardstick in psychology, with classical conditioning viewed as compelling evidence of the same.
Berlin Wall (politics)	A wall erected by the Soviets in 1961 to prevent its nationals defecting to the West. Its subsequent demolition, in 1989, marked the end of the Cold War.
Beta particles	High energy electrons, emitted by radioactive materials during beta decay.
Big bang	The dramatic decoupling of the major forces at the beginning of time, sufficient to create the expanse of space and everything within it (including time itself). In the next book, that so-called 'decoupling' is termed deep-space nucleosynthesis, and what makes it so phenomenally explosive is the absence of gravity's moderating influence.
Bilateral effects	Remote manipulation of a single person.
Binary system	A binary strings consist of a series of ones and zeros, each of which is termed a bit. Together, eight of these bits form a byte (by eight); which can represent a total of 256 values (simply by rearranging the positions of the ones and zeros).
Biosphere	That part of the earth's surface and atmosphere within which all living things are found.
Bipolar disorder	Formerly known as manic depression, this condition is characterized by intermittent mania and depression. Depressed symptoms contrast with marked mood swings; often as disparate and diverse as elation, excitability and sadness. Where such symptoms present themselves it's important that they are authentic or innate.
Black body	An imaginary object, which is a perfect absorber and emitter in all wavelengths. The light and heat emitted by a black body is wholly dependant on its temperature.

Black Power	Black Power was an ideology which arose naturally from the anger and frustration of the American civil rights movement in the 1960s. Out of this externalization of longstanding grievances emerged the Black Panther movement, which openly taunted the law with its allusion to armed resistance. However, a non-violent defence of the US Constitution proved far more effective than Black Power, as regards the empowering of African Americans.
Body politic	The whole of the electorate of a given country. In the next book, we see how social learning (as opposed to received learning) risks burdening society with unexpected outcomes – the United Kingdom's Irish Sea Border being a prime example.
Boreal zone	Climatic zone located exclusively in the northern hemisphere between the temperate zone and the arctic circle (lying approximately between latitudes 50^0 and 66^0 north). Characterized by coniferous forest.
Bourgeoisie	Refers to the middle class, which in Marxist philosophy possesses and controls capital and property, binding the working class below it to the selling of its labour.
Burqa	Islamic garment, which covers a Muslim woman from head to toe.
C	
Capitalist system	Free-market capitalism stresses that the state should be independent of economic production and wealth creation. This places responsibility for wealth creation in private hands, either through the ownership of businesses or shareholding.
Carrier wave	A radiofrequency wave onto which information is impressed through modulation. By this means radio and TV signals are transmitted.
Catalysts	A substance which speeds-up chemical reaction times, but which remains unchanged by the reaction itself.
Centre (political)	Associated with capitalism, consensus and democracy. Multi-party politics, pragmatism and public debate all dominate political life; with a strong emphasis on personal freedom, liberal individualism and civil liberties. Commonly associated with the western polyarchies.
Cerebellum	Situated within the base of the skull, this part of the brain is responsible for balance, posture and coordination.
Cerebral cortex	Conscious cognitive processing takes place in the grey matter, or cerebral cortex, which comprises the ridged exterior of each cerebral hemisphere.
Cerebral hemispheres	The left and right cerebral hemispheres are covered in gyri and sulci (ridges and furrows), with each possessing four distinct lobes – termed the frontal, parietal, temporal and occipital lobes. The surface of each consists of grey matter, beneath which lies white matter; with one of the hemispheres normally dominating.
Cerebrum	The largest part of the brain, comprising two cerebral hemispheres, each possessing four lobes (named after the cranial bones they lie adjacent to). The two hemispheres communicate via nerve fibres known as the corpus callosum. The cerebrum is the seat of learning, sensory perception, language, emotion and cognitive processing.

Chaos theory	The theory that chaos, i.e. unpredictability and disorder, can result over time from a few fundamental laws (or complex adaptive systems) behaving in a mutually-affecting manner. If those laws or systems combine to produce plasticity, fluidity and variability, then unpredictability will result. Given that such elasticity may actually be beneficial, the term chaos may not be wholly apposite.
Charge (electrical)	An aspect of the electric force which appears to exist in both a positive and negative form. According to orthodox science like charges repel, whereas opposite charges attract. The manner in which points of single charge arrange themselves in space determines much about physics and chemistry.
Chromatography	A way of isolating certain compounds and molecules, prior to their individual analysis. Chromatography exploits the fact that different molecules and compounds possess different physical and electrical characteristics.
Chromosomes	Rod-shaped structures in the nucleus of a cell containing the whole genetic code, or genes, of a given individual.
CIA	America's Central Intelligence Agency (CIA); which is responsible for intelligence and counterintelligence outside the USA.
Civil Rights Act (1964)	Landmark American legislation which prohibited discrimination on the grounds of race, colour, religion, sex or national origin. Its enactment paved the way for further legislation, most notably, the Voting Rights Act, 1965.
Classical conditioning	Based upon the ideas of Russian physiologist Ivan Pavlov (1849-1936), who realized that animals could associate two different stimuli, sufficient to condition their behaviour. This spawned numerous theories about Pavlovian conditioning in humans, and paved the way for behaviourism.
Closed system (physics)	A closed system can only exchange heat and energy with its surroundings (not mass).
Codified	Often used to describe the constitution of a country whose constitution is laid down in a single comprehensive document.
Coevolution	Also known as duel inheritance theory, coevolution suggests that social learning feeds back into the gene pool, affecting the characteristics of subsequent generations. To date, such feedback appears to have done little more than afford people superficial racial characteristics.
Cognitive dissonance	Psychological conflict arising due to incompatible thoughts, actions and beliefs. Such conflict can arise when a person's actions fall-short of their idealized self, i.e. their self-schema.
Cohort	Persons of similar age, who are likely to have shared a range of common experiences, given the time frame within which they have all lived.
Cold War	The state of political and military tension which existed between the West and the Soviet Union between 1946 and 1989.
Collectivization	Forced collectivization was common in the communist Second World, as private capital and landholdings were pooled into communes or collectives. Often the level of violence and intimidation needed to make these acts of folly materialize was available only to the most powerful demagogues, e.g. Josef Stalin and Mao Tse-tung.
Commission (Law)	Committing a criminal act (as opposed to acting criminally through omission or possession).

Common law	Laws introduced by the judiciary, often in the historical past – and which have neither been repealed, nor overtaken by the introduction of statute law.
Common Market	In 1957 the Treaty of Rome established a European Economic Community (EEC), more commonly referred to as the Common Market, in which all barriers to trade were removed between the original six members: France, Italy, Holland, Belgium, Germany and Luxemburg. The principal behind the common market remains, only the legislation and member countries has grown.
Commonwealth of Nations	The United Nations Trusteeship Council, created in 1945, ushered in a whole new era of decolonization and post-imperialism. In the case of the British Empire the move was towards an association of equal partners, termed the Commonwealth of Nations. Presently comprising 54 countries, all are bound by shared ideals, such as democracy and human rights.
Communism	Communism is a political ideology on the extreme Left of politics. Having dispossessed the bourgeoisie and aristocracy of its power, it sought (by way of a dictatorship of the proletariat) to concentrate power in the hands of a supreme leader, within a one-party system.
Communist Manifesto (1848), by Karl Marx	Co-written with Friedrich Engels (1820-95), the Communist Manifesto is principally about class struggle and "the property question". Part I: Bourgeois and Proletarians, perceives capitalism as inherently exploitative; Part II: Proletarians and Communists, supports a stateless society (but appears to provide for one-party proletarian dictatorships); Part III: Socialists and Communists, dismisses reforming alternatives; and finally, Part IV: Position of Communists in Relation to Various Opposition Parties, invites working men of all countries to unite.
Compounds	A chemical compound is formed by two or more elements, combining in definite proportions. The bonds can be either covalent or ionic.
Conative dimension	Pertaining to an individual's own will, as opposed to the will of a superimposed third party.
Congress	US legislature comprising two elected chambers: 1. the senate; and, 2. the House of Representatives. For a bill to become law it must be approved by both chambers and endorsed by the president.
Conscience theory	The argument that a conscience can be conditioned into a person, i.e. wrongdoing (conditioned stimulus) meets with punishment (unconditioned stimulus), which produces discomfort (unconditioned response), resulting in avoidance of misconduct (conditioned conscience). This results in heteronomous morality, defined principally in terms of consequences and externally-imposed controls.
Conservative Party	Right-of-centre political party, which (together with the Labour Party) dominates mainstream politics in the UK.
Constructive interference	Occurs when two identical waves are perfectly in phase, sufficient to amplify one another.

Constructivist theory	Theory proposed by the child psychologist Jean Piaget (1896-1980), in which children are seen as adopting personal constructs or schema; ones which are abandoned, re-evaluated and fine-tuned throughout the whole process of development – due to the acquisition of logical abilities, many of them taught. Many would argue that this process of personal growth continues throughout the person's life.
Continuous development	Smooth, regular and unbroken development.
Contralateral	The left hemisphere controls the right-hand side of the body and the right hemisphere the left-hand side.
Copenhagen interpretation	The prevailing view in quantum physics that experimentation can only furnish us with information about a particle's position in space, or its momentum, but not both. Contrasts with wave mechanics, which questions whether particles actually exist.
Corporatism	Corporatism is a natural extension of the fragmented elite model of democratic structure, in which large well organized elites, representing large numbers of people, have an excessive influence over (and undue access to) the executive branch of government.
Corpus	The complete body of knowledge, written or otherwise, which serves as a point of reference on a particular topic, and which therefore enriches one's appreciation of that subject.
Corpus callosum	The nerve fibres facilitating communication between the right and left cerebral hemispheres.
Cosmic inflation	The expansion of the universe by a process consistent across space and time.
Coupling	See coupling constant.
Coupling constant	Defined in this book as the amount of force required to effect a substantive change, especially in respect of such fundamental forces as the electric, magnetic and gravitational forces. Examples of such a change include light being able to radiate, ions being sufficiently attracted that they bond in an ionic manner, and a paperclip being picked-up by a U-shaped magnet. That biting point, i.e. coupling constant, betraying symmetry and conservation.
Covalent bonding	Chemical bond produced when two atoms share an electron (in order to achieve, as far as possible, full valence shells).
Cultural positivists	Those who believe that culture is that which is posited as being one's culture, whether one encounters it or not. Telepathy could be posited as being a cultural reality, but not one you personally encounter.
Cultural realists	Those who believe that culture is whatever one encounters or experiences – whether pleasant or not. This makes it an objective reflection of culture, as it's actually experienced; not the normative alternative served up cold to tourists and sightseers.
Cultural Revolution (China, 1966-76)	Mass starvation and the failure of the communes (set-up during the notorious Great Leap forward, 1958-61) resulted in a sense that the Marxist-Leninist revolution in China was about to stall, prompting the so-called Cultural Revolution. The Cultural Revolution amounted to a ideological purge of liberal-minded elements and alleged bourgeoisie.

Dark Cycle	The gravitational force, and indeed light itself, both conform to the inverse square law; such that their effects radiate outwards, becoming ever weaker. So weak, in fact, that gravity becomes gravitation, inhibiting the passage of light; whereupon any electromagnetic energy is converted into space. Space-creation forces galaxy clusters apart, amplifying this conversion of energy, sufficient to fuel cosmic inflation.
Dark energy	The energy associated with the conversion of electromagnetic radiation into the fabric of space. All energy is 'dark' without gravity to synthesize, propagate and support the same – that is to say, gravity impedes the expansion of the cosmos.
Dark matter	Objects rendered invisible due to the conversion of their electromagnetic radiation into the fabric of space.
Decision makers	In a modern liberal democracy access to mainstream power is access to the machinery of decision making, be it elected politicians, high-ranking civil servants or the executive branch of government. One could extend its meaning to include all holders of high office and those wielding significant powers.
Declaration of Independence (4 July 1776)	Written proclamation issued by the Second Continental Congress, on behalf of the American colonies, that they would henceforth govern themselves.
Decoupling	Defined in this book as a substantive change brought about by one of the fundamental forces dipping below a certain magnitude; for example, the electric, magnetic or gravitational force. The effects of decoupling include light being unable to radiate, ions being incapable of bonding, and a paperclip languishing motionless in the presence of a magnet.
Deduction (logic)	See hypothetico-deductive research.
Deformed polyarchies	Privileged access to the government and the policy-making process may give certain elites a disproportionate influence, leading in some cases to corporatism and deformed polyarchies.
Democratic Party	America's oldest political party, dating back to the 1790s, when it was called the Democratic Republican Party. After gaining office in 1801 it went from strength to strength – eventually splitting over the issue of slavery and changing its name (in 1854) to the Democratic Party.
Denial	Sceptics tend to view denial pessimistically, believing that every form of evasion is inherently wrong and self-serving. Ultimately, denial (which includes outright refutation, discrediting and renaming) represents a form of power, and like all powers it could, in theory, be put to the service or no particular individual or state.
Deoxyribonucleic acid (DNA)	DNA is the chemical basis for life, and shares some characteristics with the life it produces, e.g. it comprises mostly oxygen, carbon, hydrogen and nitrogen (elements which make up 96.6% of the human body); it has a form which combines strength with flexibility; and, crucially, it can replicate itself. DNA contains all the genetic information for a given individual (information which determines how effective that individual will be at overcoming environmental stress).

Depolarization	At rest the membrane of a neuron is slightly polarized, with the inside negatively-charged. Outside the membrane are positively-charged sodium ions, which can pass through the membrane rapidly, via sodium ion channels, when the neuron is stimulated. This sudden depolarization may be great enough to trigger a nerve impulse.
Depression	Characterized by low-mood, poor self-esteem, overwhelming pessimism and unshakable despair. Depressed individuals can be superimposed into others, most controversially when the subject is wholly unknowing of the phenomenon and its effects.
Destructive interference	Occurs when two identical waves are perfectly out of phase. Prima facie, it is akin to two identical photons neutralizing each other at a point. See also constructive interference.
Determinism	The view that all events are determined by something and that there is no free will. Biological and genetic determinism implies that one's physical makeup governs one's actions, but if one's chemistry gives rise to plasticity, elasticity and malleability then no action is wholly predetermined. More broadly, while natural systems aid prediction, through nascent patterns, emergent order and dynamic stability, despotism seeds chaos. See also, normal curve of distribution.
Devolution	Decentralization of power, such that powers are passed down to regional assemblies and local parliaments. Sovereignty, however, remains with that doing the devolving.
Differentiation (genetics)	When a spermatozoon and ovum fuse, forming a zygote, cell division produces a child possessing a complex anatomy, e.g. skeleton, muscles, central nervous system, brain, eyes, etc. Differentiation is the mechanism which switches genes on and off sufficient to account for that diversity. Processes in the cell's nucleus or cytoplasm, probably involving RNA, are likely to explain that diversification.
Differentiation (psychology)	The view that as we grow and develop we simply become more adept at differentiating between various stimuli. See also enrichment (psychology).
Diffraction	An alteration to the course of radiation, or a spreading out of the same, as a result of meeting with an edge or narrow aperture.
Digital electronic circuits	Electronic circuits which process discrete signals, either in the form of bytes or (as in the case of radio and TV signals) as so many pulses. See also binary system.
Dipolar molecules	Molecules which have a permanent separation of charge, due to the shared electrons favouring the more electronegative atom.
Dipole-dipole interactions	Intermolecular forces which exists between dipolar molecules. The region of partial positive charge on one molecule being attracted to the partial negative charge on another.
Dipole moment	Turning effect, or torque, produced by the separation of charge in dipolar molecules. Equally, the separation of charge produced by such forces (such is the nature of electromagnetism at these smallest scales). See also magnetic moment.
Discontinuous development	Intermittent, irregular and episodic development.

Disequilibrium	A state of mind in which one is unable to adequately explain events according to one's present knowledge.
Dispersion forces	Weak intermolecular forces, produced by molecules in close proximity. The electrons in one repel the electrons in the other, inducing a temporary dipole. This leads to mild electrostatic attraction.
Dispositional attribution	Arises when the cause of a particular form of behaviour is attributed to the person's personality, rather than the demands of the situation.
Dissociate	The breaking-up of a compound, such as water molecules. In the case of water, H_2O dissociates into H + OH.
Double-blind controls	A methodology used by science to avoid subjective bias invalidating a given result. Double-blind controls ensure that the experimenter is unknowing of the working hypothesis.
Due process	Due process demands that a defendant's legal rights are respected. The Magna Carta, of 1215, expresses it as follows: "no freeman shall be arrested or imprisoned or deprived of his freehold or outlawed or banished or in any way ruined... except by the lawful judgement of his equals and according to the law of the land".

E

Economic and Social Council (UN)	The United Nations' Economic and Social Council is a subordinate body of the General Assembly. Besides fostering cooperation in the area of sustainable development, it also addresses issues such as the status of women, social development and crime.
Efference copy	Repetitive learning of motor skills (i.e. actions involving movement) creates an imprint of the movement on the nervous system. According to this hypothesis, this so-called efference copy helps to make one more accomplished at the given task.
Efferent signals	Nerve impulses travelling away from the brain.
Ego	The human brain's source of consciousness, memory and planning. See also self-schema (which is near-synonymous with the ego).
Eisenhower Doctrine	US President Dwight D. Eisenhower (1890-1969) stated in 1957 that "the US wouldn't hesitate to use armed force in the event of communist aggression in the Middle east".
Elective dictatorship	Arises when the ruling political party dominates the legislative body, and the executive is constrained principally by the need to win future elections.
Electoral system	The dominant electoral system in both Britain and America is the First past the Post (FPTP) system, in which the candidate with the most votes wins. However, if there are many political parties the vote may be split in such a way that more people actually voted against the winning party. The benefit of a two-party dominated (FPTP) system is that the result is more likely to be a reflection of the people's wishes.
Electrical potentials	The potentials associated with a neuron's electrical state are 1. resting potential, 2. local potential, 3. threshold potential, and 4. action potential. The action potential is synonymous with a nerve impulse.

Electric fields	The fields surrounding charged bodies, such as points of positive or negative electric charge.
Electromagnetic induction	A moving magnetic field and a changing magnetic field can both induce electrical currents in a nearby circuit – a phenomenon known as electromagnetic induction. Radiation comprises oscillating electric and magnetic fields which induce mild currents in antennas and aerials. The equivalent of such induction exists in biological systems.
Electromagnetic radiation	Energy, emitted by a source, which disseminates according to the inverse square law. Otherwise known as simply light or radiation.
Electromagnetism	The interaction of electric and magnetic fields, sufficient to induce electrical currents and phases of matter
Electron	A particle which carries a single negative charge, but negligible mass.
Electronegativity	How strongly an atom of a given element attracts an electron.
Electronic configuration	The distribution of the electrons within the orbitals of an atom. The electronic configuration of helium is $1s^2$ (i.e. two electrons occupying an 's' orbital in shell 1).
Electron jumps	An electron either absorbs or emits a photon, and when it does the electron jumps from one orbital to another. During absorption it jumps up, and during emission it jumps down into a ground state.
Elements	Atoms of the same element contain identical numbers of protons in their nuclei. The earth has 92 naturally-occurring elements.
Elite pluralism	The state may well have a constitution which is pluralist in nature, and therefore supportive of a near infinite variety of interest groups, but in practice only certain elites are capable of strongly influencing policy.
Empirical refutation	See Falsification.
Energy crisis	Any crisis which has the effect of greatly increasing the price of petroleum (or any other vital energy resource), as happened in 1973, during the Yom Kippur War, and again in 1979, during the revolution which deposed the Shah of Iran.
Energy security	The security afforded to modern economies, through access to cheap and reliable energy sources. Conversely, energy insecurity is a major threat to global peace and stability.
Enrichment (psychology)	The view that as we grow and develop we increasingly supplement the information provided by our senses, enriching the same ever more as we get older. See also differentiation (psychology).
Environmental resistance	All species eventually meet with environmental resistance due to continental drift, volcanism, competition, climate change, asteroid impact, or numerous other factors. DNA provides for living organisms continuously readjusting to these stresses, through random mutations.
Enzymes	A specialized protein produced by living organisms for the express purpose of speeding-up biological processes.
Epiphenomenalism	Stresses the importance of physical processes; such that conscious self-awareness and the human spirit are thought of as mere side-effects of chemical reactions. Ultimately, the mind is at the mercy of the body, or should I say, chemistry.

Epistemological jurisprudence	The philosophy of law which questions what is knowable, in an age of growing telepathic awareness. For telepathy to be wisely assimilated, the proof of individual thinking should not be made available to the law, placing limits on what can be proven at a mainstream level.
Epistemology	Concerned with knowledge, particularly how we define knowledge, its demonstrable limits, and the implication of the same.
Equatorial zone	At or near the earth's equator (latitude $0°$).
Equilibrium	A state of mind in which commonly encountered information and experiences can be satisfactorily explained within the parameters of one's present knowledge.
Equivalence	$E=mc^2$ is an example of equivalence; energy being equivalent to (or equal to) mass times the speed of light squared.
European Community (EC)	The forerunner of the present European Union (which existed between 1967-1993). In 1993, with the process of ratifying the Maastricht Treaty complete, it became the European Union.
European Court of Human Rights (ECHR)	The ECHR exists due to the European Convention on Human Rights. Many countries have incorporated the latter into their legal systems; but in those cases where they haven't an EU citizen may be forced to take their case to the European Court of Human Rights.
European Court of Justice	The highest court in the European Union, as regards EU law itself, i.e. its interpretation, even-handed application, dealing with violations of the same, and addressing non-compliance.
European Monetary System (EMS)	Established by the European Community in 1979, this was an early attempt at establishing a single European currency, i.e. the European currency unit (ecu). However, the reunification of Germany in 1990, the Maastricht Treaty the following year, and an anticipated big-bang increase in countries joining the EC following the disintegration of the USSR, brought about its replacement (i.e. European Monetary Union).
European Monetary Union (EMU)	European Monetary Union was implemented over the course of three years, from 1999-2002. By the end of that time the European Union began issuing euro banknotes and coins – with a European Central Bank set-up to manage monetary policy.
European Union (EU)	Created by the Maastricht Treaty (1991), the European Union (EU) effectively swept away the European Community (EC), which was little more than a prototype, and replaced it with a growing federation.
Evolution	Organisms produce more offspring than an environment can support. As that environment changes through time, those with favourable characteristics survive. Thus, a combination of both environmental resistance and random genetic mutations gives rise to whole new species.
Executive	The executive branch of government is centred on the president in American politics and the Prime Minister (and their cabinet) in British politics. As the franchise or vote extended through time, these figures were forced to appeal to much larger and more diverse electorates. The principal function of the executive is policy-making.
Exogenous factors	Those factors arising beyond the individual; particularly factors impacting upon them directly, such as telepathically-induced effects.

245

Externalize	Problems may be externalized or internalized. Openly expressing personal grievances, in other words externalizing, may prove counterproductive (e.g. aggression, anti-social behaviour, etc).
Extrasensory	Information conveyed to one's brain other than through the five senses. Thus, remote manipulation, mental telepathy and the superimposition of persons are all extrasensory phenomena.

F

Falklands War	The Falkland Islands, a British possession located in the South Atlantic, were briefly invaded and held by Argentinean forces in 1982. A British task force soon retook the islands following several weeks of fighting.
False imprisonment	The chief reason for habeas corpus (in which restricting a person's freedom of movement must be lawful and justified). To arrest someone on the grounds of their thinking alone would be false imprisonment, for which one could be fairly compensated.
Falsification	Engagement with reality is best achieved through hypotheses which can be disproved. If not repudiated or falsified, the hypothesis can be worked up into a theory. Also called empirical refutation.
Fascism	Fascism is a political ideology on the extreme Right of politics; appearing nationalistic, militarily belligerent and frequently racist. Individuality and personal choice are abandoned, in favour of agentic conformity and unquestioning obedience to a supreme leader, within a one-party system.
Fatalism	A state of mind produced by an inability to believe in free will; the person believing instead in determinism and predestination, brought about by the oppressive actions of others. Learned helplessness and agentic conformity reflect such fatalism.
Fermat's last theorem	French mathematician Pierre de Fermat (1601-65) scribbled a note in the margin of a manuscript; namely, "it is impossible to separate…a fourth power into two fourth powers, or in general, any power higher than the second into two like powers". A mathematical proof, validating this proposition, was arrived at in 1995.
Feynman diagrams	Graphical representation of the interaction of elementary particles. Introduced by Richard P Feynman, ostensibly as an aid to understanding quantum electrodynamics. But the timing of their introduction suggests deeper political and military motives.
First World	In the aftermath of World War Two a three world typology arose, in which a decidedly pro-American capitalist First World competed with an unwaveringly communist Second World, with so many non-aligned or neutral nations looking on. Since that time the First World has grown in strength and permanency, due largely to radiological effects transcending mainstream American politics.
Fission	The splitting of an atomic nucleus, resulting in the release of energy.
Forebrain	The forebrain comprises the cerebrum and diencephalon. Within the diencephalon (beneath the thalamus) lies the hypothalamus; which contains some of the reflex centres of the autonomic nervous system. Above the diencephalon lies the cerebrum, by far the largest and most highly-developed part of the human brain.

Four Asian Tigers	The four most highly-developed economies in the far east; namely, Hong Kong, Taiwan, Singapore and South Korea. All have witnessed rapid industrialization, allied to astonishingly high growth-rates, driven by exports to some of the world's richest nations.
Fourth World	Term used to describe: 1. Indigenous groups excluded from matters of state by their own country; and, 2. the poorest and least developed of all Third World nations. The early white settler's capacity for poisoning, shooting and displacing indigenous peoples hints at unsound cogitation – ironic, given that their progeny became victims of unsound logistics, most notably on the peninsula known as Gallipoli.
Fragmented elite model	Elite pluralism, further fragmented by schisms within interest groups, serves to dilute the influence of those groups; thereby favouring the strongest and most cohesive lobbyists and organizations.
Franchise	The right to vote within a representative democracy.
Fraunhofer lines	Approximately 25,000 observed absorption lines in the spectrum of the sun, arising due to the absorption of these wavelengths by the cooler upper atmosphere.
Free protons	A nucleus of hydrogen, carrying a single positive charge. Free protons, not forming atomic nuclei, are not known to decay.
Free radicals	Oxidative processes within the body are thought to produce highly reactive atoms and molecules, possessing unpaired electrons. So reactive they damage tissue and encourage the aging process. Antioxidants are thought to reduce this damage.
Free will	The freedom to act and make choices, free of compulsion, coercion and harm. A person exercising free will is said to be acting autonomously.
Free world	The land area falling within NATO's immediate jurisdiction.
Frequency	The number of complete waves per second, emitted by a source.
Frequency modulation	A means of impressing information onto a carrier wave, by varying the wave's frequency and wavelength.
Freudian	Derived from the ideas of Sigmund Freud (1856-1939); by explicit reference to his theories, observations and writings.
Frontal lobe	The frontal lobe is the seat of consciousness, particularly the prefrontal cortex (situated behind the brow ridges). The frontal lobe also contains important motor areas, where movement is consciously initiated.
Frustration aggression hypothesis	An out-dated view of aggression, whereby violence stems from frustration. As modern history has shown there was a case for aggression when Hitler broke the terms of the Treaty of Versailles – with the failure to act resulting in unavoidable conflict.
Fundamental attribution error	Occurs when too much blame is placed on a individual's character, and too little importance given to the demands of their situation. Actor-observer bias develops this further, suggesting that similar behaviour in oneself would be attributed to the situation.
Fusion	The squeezing together of lighter elements to form heavier elements, with the balance of the combined masses being given off as light and heat.

Galaxy clusters	Galaxies are concentrated groups of stars, often forming a classic spiral shape, and which frequently exist in groups or clusters.
Gametes	Specialist cells which convey genetic information from the man and woman in sexual reproduction – spermatozoon in the case of the man and ovum in the case of the woman. Each gamete contains half the usual number of chromosomes, i.e. 23 (but when fused together into a zygote that number rises to 46).
General Assembly (UN)	The United Nations' General Assembly is the representative instrument of all the member states of the United Nations (193 at the time of writing). Each country has one vote; and when voting on particularly serious issues a two-thirds majority is required.
Generation gap	The marked difference in attitude and behaviour between one generation and the next, especially between young people and their parents.
Genes	A sequence of bases, along a strand of DNA, needed to produce a specific physical characteristic or physiological function.
Genetic mutations	Random changes to the sequence of bases (in those genes giving rise to specific physical characteristics and physiological functions) sufficient to affect the way in which the individual grows or develops.
Genocide	The deliberate extermination of an ethnic group, religious faction or race of people through mass murder.
Gestalt psychology	Gestalt means whole in German, and therefore Gestalt Psychology treats behaviour, perception and brain structure as intrinsically related – and that to focus too narrowly on any one of these obscures our deeper understanding. In the present age one must also address politics, power and multilateral telepathic involvement.
Glial cells	Specialized cells, which have evolved to support, nourish and protect the primary nerve cells of the brain, i.e. the neurons.
Globalization	The anticipated emergence of a single integrated global economy; made easier by the introduction of electronic money, satellite communication, 24 hour trading and the internet. Free-market capitalism stresses that the state should be independent of economic production and wealth creation, but that doesn't mean that a supranatural presence can't engage in neo-mercantilism in the interests of no single country.
Gravitation	Gravity and gravitation are terms applied to the gravitational force, but within this narrative the former is defined as being stronger than the latter – with the two separated by a coupling constant.
Gravity	That which binds the visible universe through mutual attraction, and which is supportive of light and radiation.
Greater Middle East	A great swathe of arid and semi-arid land straddling the line of latitude at 30^0 north. Travelling east to west one passes over Pakistan, Afghanistan, Iran, Kuwait, Iraq, Saudi Arabia, Jordan, Israel, Egypt, Libya, Algeria, and finally Morocco. To the north lies Tunisia, the Lebanon, Syria and Turkey, and to the south, Sudan, Yemen, Oman, United Arab Emirates and Qatar. Islam being the foremost religion.

Great Leap Forward (China, 1958-1961)	Mao Tse-tung's ill-fated attempt at boosting China's industrial and agricultural performance by exploiting the country's most prolific resource – labour. By concentrating the population in communist collectives or communes Mao hoped that agriculture, industry and urbanization would grow together in a productive manner. The reality proved to be mass starvation, amid agentic conformity, which could only be achieved through terror and coercion.
Groundwaves	Any wave transmitted directly to a receiving aerial, without it reflecting off the ionosphere.
Groups (chemistry)	One of the vertical columns in the periodic table of elements.
Guantanamo Bay	In October 2001, following the 9/11 attacks, a full-scale invasion of Afghanistan was launched. Hundreds of suspected al-Qaeda terrorists were apprehended by the American military as a consequence of this invasion, many of them ending-up incarcerated at Guantanamo Bay Naval Base, Cuba (in a custom-built detention facility).
H	
h	See Planck's constant.
habeas corpus (Latin: *have the body*)	An aspect of natural justice which tests whether the detention of an individual is lawful, by forcing the custodian to justify the detention to a court. It is strictly about the legitimacy of the detention, not about guilt or innocence.
Hague, The	The United Nation's International Court of Justice.
Halabja	Kurdish town which suffered a notorious gas attack in 1988 (killing 5,000 inhabitants) and which was initiated by Iraq's ruling Ba'ath Party, headed by Saddam Hussein.
H-bonding	Polar compounds, like water, have regions of pronounced positive charge in the vicinity of the hydrogen atoms. Other compounds have regions of pronounced negative charge in the vicinity of highly electronegative atoms. When these positive and negative regions stick together H-bonding occurs. H-bonding is the mechanism which provides for substances dissolving in water.
Hertz (Hz)	A measure of frequency, being the number of waves emitted by a source in just one second.
Heteronomous morality	Heteronomous morality is defined principally in terms of consequences and externally-imposed controls, having been conditioned into the individual by others.
Hidden curriculum	Term given to those unofficial codes and standards which children are expected to conform to, but which aren't formally written down.
Hindbrain	The hindbrain comprises the medulla oblongata, pons and cerebellum. This primitive part of the brain is associated with posture and motor-learning.
Holistic	The natural sciences (physics, chemistry and biology) and social sciences (sociology, psychology and economics) can be studied together in a holistic manner, so as to better understand the elasticity, plasticity and fluidity which complex interacting systems so often give rise to. The sceptic sees holism as explaining how the wider whole is rendered unknowable, through the sum of its parts – and, indeed, the science of holism might seek to determine the point at which that complex conjuncture becomes unpredictable, e.g. the future.

Holocaust	The mass extermination of European Jewry by the Nazis, during the Second World War.
Homo sapiens (**Latin:** *wise man*)	The species that is modern humans.
Human genome	*Homo sapiens* possesses 46 chromosomes arranged in pairs. The DNA contained within these chromosomes comprises some 3.2 billion base-pairings, providing for some twenty to twenty-five thousand genes. Sequencing all of these genes has been the job of the Human Genome Project (begun in 1990, and completed in 2003).
Hund's rule	This rule states that every atomic orbital in a given sub-shell must be partially filled first, prior to the addition of a second electron.
Huygens-Fresnel principle	Counterpoint to the corpuscle (particle) theory of light proposed by Sir Isaac Newton. According to this principle light behaves in the manner of a wave, with its predicted effects adding weight to that theory.
Hydrophilic	Water-loving; readily forming H-bonds with water.
Hydrophobic	Water-hating, avoiding interacting with water.
Hyperpolarization	Exceptionally, the photosensitive pigments in the eye absorb incident radiation triggering an initial hyperpolarization in the adjacent sensory neuron, rather than depolarization.
Hypothesis	A statement which can be tested by experiment and observation, often predicting the result; and which is capable of being falsified or rebutted by the said tests and observations.
Hypothetico-deductive research	Deductive research undertaken by way of the scientific method, i.e. a refutable hypothesis is tested by experiment. These hypotheses often conform to prevailing theory, leading in turn to a compelling paradigm.
I	
Id	The human brain's unconscious source of primitive drives and desires.
Inadmissible (Law)	Evidence which is inadmissible cannot form part of formal legal proceedings. Hearsay, in which a person conveys the testimony of another (who isn't available to be cross-examined), is an example of inadmissible evidence. It is proposed to make comments regarding another person's thoughts inadmissible.
INAH3	The interstitial nucleus of the anterior hypothalamus (INAH3) is significantly larger in men than women and is known to influence and affect sexual orientation and behaviour in mammals.
Incommensurability	What happens when tests and observations produce results at odds with current or accepted theory, resulting in the need for a revision of the same.
Indirect rule	Occurs due to the instruments of the state being co-opted by an external or overarching power, via radiofrequency technologies.
Induced dipole	Occurs when two molecules are in close proximity, and the electrons in one repel the electrons in the other, producing a temporary dipole. These two molecules then experience a mild electrostatic attraction.

Induction (logic)	With induction observation comes first; from which theory is then derived. As opposed to science, in which the hypothesis comes first; whereupon tests and observations follow.
Innate characteristics	Personal characteristics arising within a given individual, rather than via some external source.
Interference	What happens when two waves of the same frequency, wavelength and amplitude meet. The effect can be constructive or destructive.
Intergovernmental Panel on Climate Change (IPCC)	Established by the United Nation's Environment Programme (UNEP) and the World Meteorological Organization (WMO) in 1988, the IPCC critically assesses the current data regarding global climate change.
Intermanual conflict	So-called 'alien hand syndrome' could arise due to the impediment of inhibitory signals within the brain, or as a result of the direct superimposition of motor impulses.
Intermolecular forces	Forces arising between molecules. Within biological systems such forces comprise dipole-dipole interactions, H-bonding, dispersion forces and salt bridges.
Internalize	Problems may be internalized, rather than externalized, resulting in protracted thinking about the said issues. Such thinking may result in a disturbance of mood (e.g. worry, anxiety, guilt, depression, etc).
International Atomic Energy Agency (IAEA)	Independent intergovernmental organization, allied to the UN, which promotes the peaceful use of nuclear energy and ensures that states comply with the nuclear Non-Proliferation Treaty (NPT).
International Court of Justice (UN)	The United Nations' International Court of Justice is the foremost world court and the main judicial instrument of the UN. Its 15 judges are elected by the General assembly and the Security Council. Its power is heavily reliant on consent.
International Energy Agency (IEA)	The IEA has twenty-eight member countries; which reads like a list of NATO members and affiliates. The IEA is primarily concerned with energy security, economic development, environmental awareness and global engagement. Through its work, NATO is able to anticipate and avoid problems arising from energy insecurity. Looking ahead, it enables NATO to plan for the post-peak future.
International law	The law governing the actions of states, and which comprises so many conventions, treaties and accepted codes of conduct.
Intifada	Several Palestinian uprisings in the West Bank and Gaza Strip, beginning in 1987, in protest at Israeli expansion and hegemony.
Intramolecular forces	Forces found within molecules. Within biological systems that is almost invariably those forces which provide for covalent bonding.
Inverse square law	Light intensity is proportional to the square of the distance from the source; so that as the distance doubles its strength is reduced to a quarter of its original value. This is common to other phenomena, such as gravity, magnetism, and the forces exerted by an electric field.
Ion	An atom which has become electrically charged due to gaining or losing electrons.
Ion channels	See sodium ion channels.

Ionic bonding	Occurs when one atom donates an electron to another atom, creating a strong electrostatic attraction between the two.
Ionizing radiation	Gamma rays, X-rays and ultraviolet radiation can break chemical bonds and remove electrons from their parent atoms.
IRA	Paramilitary organization dating back to the early 20th century, which was dedicated first to the expulsion of the British from mainland Ireland, and then to the reunification of Ireland following partition.
Isolated system (physics)	A system which is unable to exchange anything with its surroundings (neither mass, energy or heat).
Isomerism	Molecules possessing the same number and types of atoms, but arranged in a different way. Isomerism produces an alternative structure, which possesses different properties.
Isotopes	Atoms of the same chemical element, possessing identical numbers of protons, but different numbers of neutrons.
Isotropic medium	Having properties which are uniform in all directions.
J	
Judiciary	The principal role of the judiciary is to decide on legal cases, arising due to laws being allegedly transgressed. How clear and well-defined those laws are determines how much the judiciary is called upon to interpret the same.
Jurisprudence	The theory, science and philosophy of the law.
K	
Khmer Rouge	A communist guerilla army, led by Pol Pot, which gained control of Cambodia in 1975, whereupon it forced the inhabitants of the capital Phnom Penh out into the killing fields. As many as two million people were murdered by the Khmer Rouge prior to the country's invasion by Vietnamese forces in 1979.
Korean War (1950-53)	The Korean Peninsula was partitioned following the conclusion of World War Two. Forces from communist North Korea invaded the south in 1950, resulting in a UN sponsored military retaliation from the pro-American south. The US commitment in Western Europe has been proposed as a possible explanation for the reluctance to use maximum force on the part of the Americans. The effect was a difficult stalemate, one we still live with today.
Ku Klux Klan	White supremacist reactionary movement which arose in the US following the American Civil War (1861-65). Ironically, it gained momentum throughout the progressive era (1890-1930), resulting in the lynching and shooting of perceived undesirables. Membership of the organization crashed from the 1940s onwards.
L	
Labour Party	Left-of-centre political party, which (together with the Conservative Party) dominates mainstream politics in the UK.
Left (political)	The political Left has typically been associated with nationalized industries within otherwise democratic capitalist economies. More radical was communism, including Marxist-Leninism, which held that both capitalism and democracy should give way to a one-party system, amounting to a dictatorship of the proletariat.

Legal positivism	The view that the law is simply what is posited, often for reasons which are inherently utilitarian and pragmatic. This contrasts with natural law which views the law as a reflection of general morality, rather than as a practical device or convenient instrument.
Legal realism	The view that the law is what you get in a given place at a given time, given all the factors present. This is an objective appraisal of what the law is actually doing; as opposed to what it ought to be doing, which is a normative question.
Legislature	Legislatures are assemblies whose central role is the scrutiny of laws and policies which the executive branch of government is keen to introduce. Often comprising representatives from several political parties they ensure that the laws which are passed are ostensibly a reflection of the peoples own will.
Legitimate power	In general terms, power that is wielded in a socially acceptable manner by persons of suitable standing. More specifically, in the context of this book, power which is applied in a well-balanced manner, sufficient to define reasonable limits, whilst providing for a suitably rewarding mainstream social environment. See also Centre (political).
Lib Dems	The decline of the Liberal Party from the First World War onwards led to a two-party system for much of the remaining 20th century. Eventually, the Labour Party produced a disgruntled faction, which broke away and formed the Social Democratic Party (SDP); which allied itself to the ailing Liberal Party. In 1988 these two parties merged adopting the name the Lib Dems (short for Liberal Democrats).
Liberal individualism	The West's much vaunted commitment to personal freedom; one which sees the rights of the individual as being of primary importance in all things political. An important commitment, if only as a defence against the corrosion of civil liberties.
Liberal Party	See Lib Dems.
Libertarian	The belief that free will is both possible and desirable in modern societies. Free will being defined as thought and behaviour which is autonomously arrived at, free of physical interference by third parties. Leaving aside the question of inherited character and social conditioning, autonomy means non-interference by others.
Libido	Analytical psychologist Carl Gustav Jung (1875-1961) saw libido as more than simply the sex-drive, but rather a psychical energy incorporating the need for social equilibrium – disequilibrium resulting from an inability to account for feelings, experiences and events.
Light	Colloquial or informal term for electromagnetic radiation.
Limited effects theorists	A theory of mass communication which argues that individual preferences strongly dictate how we access and perceive information provided by the media, thereby greatly limiting its influence.
Lipid bilayers	Each lipid molecule possesses a water-loving and water-hating region. When these molecules form into a cell membrane the water-loving hydrophilic parts are exposed to the aqueous environment of the cell and surrounding tissue fluid. Whereas the water-hating, hydrophobic parts, are shielded from the same.
Local potential	That level of depolarization of a neuron's membrane which is insufficient to trigger a nerve impulse; and which will result in the nerve cell returning immediately to its resting state if not exceeded.

253

Macrosociology	The sociological study of society as a whole.
Magnetic configuration	The strength, distribution and influence of all the magnetic fields, in and around an atom or molecule. This configuration is hypothesized to affect absorption and emission, as regards the said molecule or atom.
Magnetic fields	The field surrounding a magnet, moving electric charge, or current-carrying coil.
Magnetic moments	Turning effect, or torque, produced by magnetic forces arising in dipolar molecules. For an atom, in such a molecule, to attract to itself more negative charges than there are positive charges, one requires this additional force, or moment. This moment being the sum total of all those additions and subtraction to the magnetic field.
Magnetic resonance imaging (MRI)	When the human body is placed in a strong magnetic field the body's soft tissues emit signals which can be detected, processed and converted into images by computer.
Mainstream	The socio-political environment comprising both the national and subnational levels; together with the institutions they contain and commonly deal with. Contrasts with NEURON, which would occupy a position transcending the whole of mainstream society.
Mainstream politics	Politics as commonly encountered at the national and subnational levels, and between nation-states. Punishment is best left to the mainstream political environment – leaving a supranational presence, i.e. NEURON, to reward those well-balanced individuals who respect the core principles of natural justice
Maoism	Invariably in extreme Left-wing one-party dictatorships Marxist-Leninism is re-shaped to fit the whims and caprices of its central demagogue. Thus we arrive at Stalinism (Josef Stalin), Titoism (Josip Broz Tito) and Maoism (Mao Tse-tung).
Mass	The gravitational mass of an object relative to another, i.e. the strength of attraction at the surface of a celestial body sufficient to give rise to weight. Such weight creates inertia or resistance to changes in the object's motion or movement.
Mass spectrometry	A technique which measures the amount of force required to deflect a vaporized sample of a given substance. The amount of force required being an indication of the mass of the particles.
Materialism	Emphasizes the importance of the inanimate, from which everything derives, not least of which, the animate.
Mecca	The birthplace of the prophet Mohammed (c.570-c.632). Located in western Saudi Arabia, this is Islam's most sacred site.
Mechanical determinism	This is the ineluctable conclusion which epiphenomenalists, atomists, materialists, physicalists and behaviourists are drawn to, but which presumes that chemistry and physics don't create a reality fraught with malleability, adaptability and unpredictability.
Medina	Site of the tomb of the prophet Mohammed. Located in western Saudi Arabia, this is Islam's second holiest site, second only to Mecca.

MEG	Magnetoencephalography (MEG) measures small localized changes to the magnetic fields produced by the outermost regions of the brain. Twenty-five thousand Fraunhofer lines in the spectrum of the sun, and only 92 naturally-occurring elements, proves that these magnetic fluctuations are likely to affect only a minute fraction of the brain's entire absorption and emission spectrum.
Memory	One of humankind's most important faculties, and also one of the least understood. Our ability to store away ideas, recollections and stories has served our species immensely – but now the semi-isolated environment of our minds has been opened-up to wider scrutiny and perusal, often by people or persons unknown.
mens rea (Latin: the guilty mind)	The state of mind which must be present for an act to be deemed a crime (whether it be an act of commission, omission or possession). For example, diminished responsibility or an epileptic fit may prevent an act being classified as a crime.
Mental telepathy	Transfer of thoughts and cognition, by way of radiofrequency technologies. A multilateral approach would involve several persons.
Microsociology	The sociological study of small groups and individuals.
Midbrain	Controls the visual and auditory systems, and contains the reticular formation which filters sensory signals to the forebrain. This is also the region of the brain which places a person in a state of sleep.
Middle East	The region to the east of the Mediterranean, comprising a number of oil-rich Arab states and which is home to Islam's holiest sites. Politics in the region is dominated by the Arab-Israeli conflict (which dates back to 1948).
Military bloc	States which forge political and military ties in order to deter potential aggressors and as a means of combining their military capabilities in the event of hostilities. NATO is presently the most powerful of all these blocs, with a potential which has hitherto remained hidden.
Modality	A category of sensation, and thus related to the five senses, i.e. touch, taste, hearing, smell and vision. Telepathy is arguably a sixth modality, but extrasensory.
Modulation	Modulation impresses information onto a carrier wave by altering its fundamental characteristics. In this way, information, dialogue and music is broadcast to those tuned into that particular carrier wave.
Molecules	A number of atoms chemically bonded together.
Mujahedin	Islamic guerillas who fought a holy war against the Soviets in Afghanistan during the Russian occupation of the country (1979-89).
Multilateral effects	Remote integration of multiple individuals, sufficient to facilitate mental telepathy, superimposition and psychokinesis.
Mutually assured destruction (MAD)	Considered to have been the likely outcome of full-blown nuclear war between the West and Soviet Union during the Cold War era. The insipient collapse of the Soviet Union, from the late 1970s, left those of school-leaving age vulnerable to alarming levels of remote interference – resulting in a more compliant cohort, in the event of a ground-war in Europe.

255

Nationalist/s	A person who subscribes to the doctrine of nationalism, which insists that the state should be free of external power and influence. Extreme nationalists will serve their country through belligerence, warring and aggression.
Nation-building	The establishment of representative democracy in place of despotism, wherever and whenever the situation demands it.
NATO	The North Atlantic Treaty Organization (NATO) was formed as a result of the North Atlantic Treaty (1949). The signatories all reaffirmed their faith in the purposes of the United Nations and their desire to live in peace with all peoples and all governments; with the parties separately and jointly, by means of continuous and effective self-help and mutual aid, vowing to maintain and develop their individual and collective capacity to resist armed attack. Presently comprises the USA, UK, Europe and all their allies. See also Free World.
Natural culture	Culture at its most practical and commonplace, e.g. driving on either the left or right, text-messaging and the free-market economy.
Natural justice	Those principles which the law must adhere to if justice is to be seen to be done. For example, natural justice implies that a verdict cannot stand if the defendant wasn't given a reasonable opportunity to state their case, and to know and answer the other side's case. Natural justice is heavily corroded by social media, television and the internet, i.e. they all pervert the core principle that 'no one should be condemned unheard' (and, that 'legal proceedings ought not to be biased').
Natural law	The view that the law is a reflection of our natural instincts, as regards general abhorrence and disapproval, and that the morality underlying that disapproval transcends all social and cultural boundaries.
Nature (psychology)	That aspect of the human psyche which derives from the chemistry and fabric of the brain, rather than what is impressed onto the same.
Nature (reality)	Colloquial term for reality, as studied by science.
Neo-Darwinian synthesis	Naturalist Charles Darwin (1809-82) published On the Origin of Species by Means of Natural selection in 1859. Shortly afterwards, in 1866, botanist Gregor Mendel (1822-84) established the principles of inheritance, through his work on plant hybrids. The subsequent discovery of the structure of DNA by geneticist James Watson (1928-) and molecular biologist Francis Crick (1916-2004) further added to our understanding of evolution. So much has been added to Darwin's initial understanding of evolution that the term neo-Darwinian synthesis has been coined to explain that most fundamental of mechanisms.
Neo-mercantilism	Manipulation of commerce and trade through the application of neuro-cognitive effects. This might afford legitimate powers commercial advantages, or advantage the global economy in all our interests.
Neural pathways	Neurons work together in unison, via neural pathways, thus providing for the transmission of nerve impulses to and from various parts of the body. The nature and complexity of these pathways is more important than the actual number of neurons. Such pathways are said to converge, diverge and reverberate.

Neuro-cognitive power (or effect)	Powers and effects derived from radiofrequency technologies, and which draw on the human brain's naturally-extant telepathic potential. Only legitimate powers should be permitted to capitalize on this neuro-cognitive capability. Ultimate control rests with the political Centre.
Neuro-linguistic programming	The marrying together of psychology, language and behaviour, in a proactive way – so as to shape individual perception for the better.
Neurological switching	The conversion of radiofrequency signals (derived from one individual's brain) into signals which can be transmitted to another's brain.
Neurology	The branch of medicine which deals with the central and peripheral nervous systems (comprising the brain, spinal cord, and associated nerves).
NEURON (politics)	Organization within which neuro-cognitive power and influence is concentrated, making it the sole provider of telepathically-induced effects worldwide. It would be the present phenomenon in its revised form – the official language of the same being English.
Neurons	Nerve cells, evolved to convey impulses to and from various parts of the body, and the cells directly involved in brain function.
Neurosis	Catch-all term for a variety of disorders, ranging from anxiety symptoms to obsessive compulsive disorders, and from hypochondria to general phobias. The person's confidence and self-esteem are likely to suffer due to the inhibiting effects of neurotic behaviour.
Neurospace	The common meeting ground produced by multilateral telepathic involvement; begging questions related to privacy and personal space.
Neurotransmission	The chemical transmission of a nerve impulse across a synapse, allowing that signal to pass from one neuron to another.
Neurotransmitters	Chemical substances which facilitate or inhibit the passage of signals across synapses.
Neutron	A particle which carries no charge, but possesses the same mass as a proton (which carries a positive charge). Free neutrons are thought to decay into protons and electrons. Together, neutrons and protons form the nuclei of atoms.
New Labour	The Labour Party, troubled by a "winter of discontent", spiralling inflation (for which a lurch to the right appeared to be the only answer), and internal discord (arising due to militant elements within), took the bold step of reinventing itself as a slightly Left-of-centre party. In this new guise, under the leadership of Tony Blair, it won a landslide victory in 1997. See also Labour Party.
Nitrogenous base	Each single strand of DNA consists of vast numbers of nucleotides, each of which has a nitrogenous base. These nitrogenous bases come in four varieties, i.e. guanine (G), cytosine (C), adenine (A), and thymine (T). However, when two complimentary strands of DNA are bound together into a double-helix, via their nitrogenous bases, only the following base-pairings are possible, A-T and C-G.
Nobel Prize	An illustrious medal awarded every year in the fields of physics, chemistry, economics, medicine and peace. This award, which was established by the will of the renowned chemist and industrialist Alfred Nobel (1833-96), dates back to 1901.

Normal curve of distribution	A symmetrical bell-shaped curve arises in many statistical analyses of biological systems, whenever a given variable is plotted against the number of occurrences. Its significance lies in the predictions it enables scientists to make when studying those systems.
Normal science	The prevailing paradigm or conceptual framework provides for normal science, which is limited in scale, incremental and one-directional (the effects of which are unlikely to be felt beyond a given branch, field or discipline).
Normative analysis	Ascertaining how things ought to be, in an idealized way.
Not self (surgery)	The surgical term for the way in which the body's immune system is able to distinguish between one's own body and the cells of another. Another's cells and organs (not self) are attacked by the immune system, leading to the destruction or rejection of transplanted organs. A cancer cure may involve the induced rejection of such a growth.
Nuclear non-Proliferation Treaty (NPT)	Historically significant treaty, which came into force in 1970, and which has sought ever since to limit the number of countries possessing nuclear weapons to just five – namely, the five permanent members of the UN Security Council. Other countries are thought to have appropriated nuclear weapons technology, presumably via those five permanent members.
Nucleotides	These are the discrete monomers (small, near-identical molecules) which join to form nucleic acids. Nucleotides have three essential components: a nitrogenous base; a sugar; and a phosphate group.
Nucleus (atom)	The minute positively-charged nucleus of an atom comprises both neutrons and protons, held together by the strong nuclear force.
Nucleus (biology)	That part of the cell which contains the chromosomes, and hence the genetic code for the individual.
Nurture (psychology)	That aspect of the human psyche which derives from what is impressed onto the brain through learning and experience, as opposed to that which is derived from the fabric and chemistry of the brain itself.
O	
Objective analysis	Ascertaining how things are in reality, however imperfect.
Ockham's razor	The view that the simplest explanation which fits the observed facts is to be preferred. However, simple induction could lead to a false conclusion. Far better that this razor is used to produce hypotheses – from which further tests and observations then follow.
Omission (Law)	A crime by virtue of omitting to do something (as opposed to acting criminally through commission or possession).
OPEC	The Organization of the Petroleum Exporting Countries exists to coordinate and unify the policies of its members; and in order to ensure a stable oil market and an uninterrupted supply of petroleum to global consumers.
Open system (physics)	An open system is able to exchange heat, energy and mass with its surroundings.
Orbitals (atoms)	Each atomic shell contains a specified number of orbitals; with each orbital capable of holding two electrons.
Organic chemistry	Chemistry associated with carbon and usually biological processes.

258

Paradigm	The conceptual framework within which scientific research is typically carried out. Corresponds with normal science.
Paradigm shift	A revolutionary repositioning of science, often due to the irrefutable proof derived from previously unheard of technologies.
Parliament	In a parliamentary monarchy parliament is the supreme legislative body. In the UK parliament comprises both the House of Commons and the House of Lords.
Particle physics	The view that whilst energy is often conveyed as waves, it interacts with matter, and indeed other particles, as a particle.
Pauli exclusion principle	States that no more than two electrons can occupy a given orbital.
Peak theory	This theory proposes that naturally-occurring resources (e.g. petroleum, natural gas, coal and uranium) will be subject to a peak in production; whereupon the prognosis will be one of terminal decline, as regards extraction, refinement and supply. According to contemporary estimates we are likely to witness many such peaks in the 21st century, due to rising demand. Any weakening of NATO (by disgruntled ultranationalists, Brexiteers and Europhobes) is to be discouraged, as the post-peak future is likely to be one of terminal decline, aggravated by conflict.
Pearl Harbour (7 Dec 1941)	A US naval base in the pacific ocean state of Hawaii, which suffered a surprise attack by the Japanese on 7 December 1941. Aerial bombing by the Japanese deprived the Americans of many battleships and aircraft, but crucially left their oil storage facility intact.
People's Republic of China (PRC)	Mao Tse-tung declared mainland China to be the communist People's Republic of China in 1949, following a bitterly-fought civil war. His nationalist opponents, led by Chiang Kai-Shek, retreated to the island of Taiwan, where they established the Republic of China (ROC).
Periodic table	A table of all the chemical elements arranged according to their atomic numbers. This table comprises rows (called periods) and columns (called groups).
Periods (chemistry)	One of the horizontal rows in the periodic table of elements.
Personal constructs	Constructs or schema which the individual appropriates, develops or invents, which are then subject to tests and observations. If found wanting new schema are called for, otherwise the existing constructs are further refined. See also schema.
Phenomenon	Clandestine power which has exploited the naturally-extant telepathic potential within every human brain. The term phenomenon specifically refers to that time period prior to full public knowledge of these neuro-cognitive effects. Once made public and reformed, however, it is proposed that it be renamed NEURON.
Phobias	Acute fear arising in response to stimuli not generally considered excessively threatening, e.g. open-spaces, spiders, heights, etc. Some may prove phobic as regards telepathically-induced effects.
Phosphate group	One of three component parts of a nucleotide, essential to the overall structure of DNA.

Phosphodiester bonds	These are the covalent bonds which join nucleotides together into linear strands of DNA. Each nucleotide comprises a nitrogenous base, a sugar, and a phosphate group – the sugar of one nucleotide is bonded to the phosphate group of the adjacent nucleotide (with this pattern repeated along the length of the DNA strand).
Photoelectric effect	The release of electrons from a surface when exposed to incident radiation of a certain wavelength.
Photoisomerization	Isomers are molecules possessing the same number and types of atoms, but arranged in a different way. Photoisomerization utilizes light energy to produce the alternative structure, which then possesses different properties and/or triggers certain events.
Photon	An elementary quantum of action, consistent with electromagnetic radiation possessing the properties of a particle, as well as a wave.
Photoreceptor cells	A receptor cell which is specifically light sensitive, e.g. the rods and cones of the vertebrate eye.
Physicalism	Philosopher Rene Descartes (1596-1650) perceived all truth in science as something which could be expressed mathematically. Such mathematics pertains largely to physical events, making physical events and mathematics arguably the only meaningful yardsticks, scientifically.
Planck's constant (h)	Has the value 6.626×10^{-34} js. Thus, the energy in a quantum of light is given by $E = hv$ (where v is the frequency).
Planned serendipity	A carefully managed outcome, arrived at through the use of radiofrequency effects, which appears on the face of it to be simply fortuitous. An example would be the distancing of Stalin and Tito.
Plasma membrane	Membrane enclosing a cell. See also lipid bilayers.
Plasticity	The malleability of a person's character, throughout the process of development, when subject to inculcation, experience and stress.
Pluralist model	Pluralism implies that the state, its instruments, and the constitution, exist to serve a potentially infinite variety of interest groups. In much the same way that an overarching neuro-cognitive power would be capable of supporting a plurality of often self-interested nations.
Polar zone	Climatic zone situated within the arctic and Antarctic circles (above and below 66^0 north and south).
Polymers	Large molecules made of smaller building blocks chemically bonded together. Those building blocks are called monomers (in the case of DNA those monomers have a specific name, i.e. nucleotides).
Polymorphism	Occurs when a species has varieties which vary solely in appearance. In the case of humans, man-made climates add to the effects of geography and climate, but in equally superficial ways.
Pons	Located in the hindbrain, the pons serves as an important junction as regards nerve fibres from various regions of the brain and spinal cord.
Pop Art	A highly popular late 20th century artistic movement which borrowed heavily from low-art, popular culture and advertising, to create bold creative statements.
Possession (Law)	A crime by virtue of possessing something (as opposed to acting criminally through omission or commission).

Postmodernism	The modern world arose in the shadow of diverse politics, spanning the range from the extreme Right to radical Left. Postmodernism implies that the political Centre now dominates, providing for greater freedom of expression.
PRC	See People's Republic of China.
Predestination (philosophy)	Fatalist's subscribe to a sense of predestination, in which free will is replaced by compulsion, oppression and third party interference. Predestination implies individual powerlessness, with events very much determined by the actions of others.
Prefrontal cortex (neurology)	The region of the brain associated with conscious thought and inhibitory signals. If the id resides in the hypothalamus (fight or flight, rage, etc), and the ego in the cerebral hemispheres, then the superego is located in the prefrontal cortex. Between them they manufacture a self schema; which hypothalamic reflexes, driven by hormones, might willfully unravel. See also cognitive dissonance.
Prescriptive rules	Recommended practices.
President	The elected executive figure in a republic, responsible for policy formation and for ceremonial and diplomatic duties overseas.
Presumption of innocence	Whilst the police often presume guilt, they must never be guilty of subjective bias in the inquiries they undertake. Such evidence accruing from those inquiries can then be put to a judge and jury (operating the presumption of innocence). For the presumption of innocence to be fairly rebutted requires two things; firstly, unbiased police inquiry; and, secondly, sufficient evidence.
Prevention of Terrorism Acts (1974-89)	A series of legislative measures introduced in the UK between 1974 and 1989, due to the rapid resurgence in sectarian violence in Northern Ireland at that time. These acts were seen as strictly temporary measures, quite unlike the growing wave of anti-terrorist legislation arising in the wake of 9/11.
Prima facie	At first glance, or judged in a cursory manner.
Prime Minister	The elected executive figure in a parliamentary system, responsible for policy formation and for diplomatic duties overseas (ceremonial duties often being undertaken by the monarch in a parliamentary democracy).
Progressive era	Period in American social and political history (circa. 1890-1930) when America witnessed increased immigration, heightened industrial output, women's suffrage, electrification, radio-communication, the birth of Hollywood, the ubiquitous car culture, jazz, and the race to realize neuro-cognitive effects in secret.
Proletariat	Refers to the working class, which is forced to sell its labour to meet the basic requirements of life, without the security of capital.
Proscriptive rules	Things to avoid.
Proton	A particle which carries a single positive charge, and the same mass of a neutron (which carries no charge at all). Together, protons and neutrons form the nuclei of atoms.
Psyche	In Freudian terms the psyche (or self) comprises an id (which is the unconscious source of primitive drives and desires), an ego (which provides for consciousness, memory and planning), and a superego (that part of the mind which acts as a conscience to the ego). The triune brain, an idea introduced by American neuroscientist Paul MacLean (1913-2007), effectively mirrors the Freudian mind, i.e. reptilian complex (id), paleomammalian complex (ego) and neomammalian complex (superego).

Psychokinesis	That aspect of superimposition specifically concerned with movement, and which (in isolation) results in an effect similar to intermanual conflict. In sport it could be performance enhancing.
Psychosis	Symptoms of psychosis include hallucinations, delusions, distorted perception and a loss of contact with reality. But many otherwise healthy people harbour delusions or are manipulated by heavily mediated messages. The question is whether one is maladaptive.
Punishment	Used to define the limits of acceptable behaviour. There are many reasons (social, psychological and economic) for setting reasonable limits and using only as much punishment as is needed to define the same. See also, note 54

Q

Quantum electrodynamics (QED)	The postwar study of the interaction of light and matter. Or, more specifically, the interaction of electrically neutral photons and negatively-charged electrons. The writer perceives QED as the appropriation of the subject by the political and military powers.
Quantum numbers	A set of numbers describing the quantum state of an electron; three of which define its position in three-dimensional space, and a fourth which pertains to spin (of which there are two kinds). This limits the number of electrons which can occupy an orbital to two.
Quantum of action	The indivisible amount of energy associated with a given photon. The relationship between the energy in a quantum of light and its frequency is determined using Planck's constant (h). See Planck's constant.

R

Radar	Radio detection and ranging was developed in the 1930s. Because objects reflect certain frequencies, the time taken for those signals to return to those transmitting the same gives an indication of the object's position.
Radiation	Energy, emitted by a source, which disseminates according to the inverse square law. Otherwise known as electromagnetic radiation.
Radiation pressure	Pressure exerted on a surface by incident radiation. It is due to the rate of transfer of the wave and the momentum of its constituent particles.
Radiofrequency attribution	Where the cause of events is attributed to remote interference via radiofrequency technologies. If one combines this with actor-observer bias, then one is more likely to attribute negative behaviour in oneself to radiofrequency effects, and the same behaviour in others to their character.
Radiofrequency technologies	Technologies capable of directly influencing electrical activity in the brain, by way of radiofrequency transmissions. However, the term can be stretched to include other wavelengths (e.g. extra-low frequencies). Such technologies exploit the brain's naturally-extant telepathic potential, born of its person-specific absorption profile.
Reactants	Substances which chemically react, producing products.
Reactivity	The likelihood of an atom or molecule taking part in a reaction. Some atoms, like oxygen, are highly reactive.

Received Pronunciation	The prescriptive practices and proscriptive practices when speaking a given language, i.e. how best to speak and what forms of enunciation to avoid. In the free world received pronunciation has become something of an anachronism.
Receptor cells	Specialist nerve cells which form part of the peripheral nervous system, and which are sensitive to touch, taste, smell, sound and light.
Reciprocal	The reciprocal of a is the number 1 divided by a (a^{-1}) See also wavenumber.
Reductionist	Pertaining to reductionism within the natural sciences (physics, chemistry and biology) and social sciences (sociology, psychology and economics). By breaking these branches down into their smallest divisible parts an understanding of fundamental laws, elementary processes and individual interactions is arrived at.
Reference power	The power which accrues to a person by virtue of affiliation to an organization, institution or body, and which has the effect of amplifying ones own power and influence.
Reflection	An alteration to the direction of radiation when it strikes a surface.
Refraction	Effect caused by radiation passing from one medium to another, whereupon its direction is affected.
Reinforcement theory	A theory of mass communication which argues that one's perceptions are simply reinforced by the information provided by the media.
Relative atomic mass	The mass of an atom of a given element. An element's relative atomic mass, measured in atomic mass units (amu), is a weighted average of all the naturally-occurring isotopes.
Remote manipulation	The direct manipulation of one person by another, by means of radiofrequency technologies. Effects may be physical, physiological and cognitive.
Republican Party	America's Republican Party arose out of the contentious issue of slavery in the mid-19th century. Those within the Democratic Republican Party who were anti-slavery broke away in 1854 and formed the Republican Party. The Republican Party has done much to preserve the union and strengthen the hand of federal government.
Republic of China (ROC)	The Republic of China was ousted from mainland China by communists under Mao Tse-tung in 1949. Thus, the Republic of China, which had held China's seat at the UN (and which had helped to establish the UN initially) found itself displaced by communists. Communists who then assumed China's seat at the UN in 1971. The ROC presently occupies the island of Taiwan.
Resilience	The ability of a person to recover quickly from adverse effects, or the tenacity to continue in spite of negative experiences.
Resistance (electrical)	Electrical resistance, by which we mean the tendency of a material to resist the flow of an electrical current, is a reflection of what the material is made from, its length, thickness and temperature.
Respiration	Can refer to the exchange of carbon dioxide and oxygen between the organism and its environment, or to aerobic respiration within cells.

Retinal molecule	A molecule which changes its shape, in a manner known as isomerism, in response to the absorption of visible light by an adjacent protein molecule. Therein rests the whole of vision in humans.
Reuters	Major news agency (based in London, UK) which acts as a primary source of news material for newspapers, TV companies and radio stations globally.
Right (politics)	The political Right has typically been associated with the capital-owning upper-middle classes, patriotism and military strength. In its more extreme form it merges with fascism, becoming overtly nationalistic, militarily belligerent and racist.
RNA	A triplet of base-pairings along a length of DNA contains enough genetic information to code for one of twelve amino acids (the other eight amino acids have to be ingested). When the DNA double-helix 'unzips', ribonucleic acid (RNA) makes a copy of the protein sequence (transcription) and conveys that code to a body in the cytoplasm, called a ribosome, where the protein is then constructed (translation).
ROC	See Republic of China.
Rule, The	Favour reward; use punishment sparingly, and only to define the limits. Those who abide by the rule view a person's 'god-given' revulsion at Anglicanism's diabolical roots as more plausible evidence of benevolence than any amount of injudicious positive-reinforcement of its protestant liturgy.
Russia	An informal term for those lands forming imperial Russia up to 1917; and for those lands comprising the whole of the USSR between 1917 and 1991. Since 1991 the term has been synonymous with the Russian federation. See also Russians.
Russian Federation	Covering 75% of the land area of the former USSR, the Russian Federation continues to dominate much of eastern Europe and northern Asia. Today, Russia has a multi-party political system, with executive power split between President and Prime Minister, a two chamber legislature, plus a Ministry of Justice.
Russians	A colloquial or informal term for the those wielding power in Moscow, and often applied to those acting as their agents in lands falling under their direct influence. See also Russia.
S	
Sakharov Prize	The Sakharov Prize for Freedom of Thought is awarded in honour of the Soviet dissident and nuclear physicist Andrei Sakharov (1921-89). A passionate advocate of civil liberties, political reform and informed public opinion, Sakharov exhibited a libertarian conscience amid so much agentic conformity and learned helplessness.
Salt-bridges	Amino acids, from which proteins are formed, often have side-chains possessing regions of full positive and negative charge. Ionic forces are then able bring these regions together, thereby adding to the 3-dimensional structure and overall strength of the resultant proteins.
Sceptic	A person who believes that we can never really know the truth of things, either due to the inherent failings of various abstract models, the illusionary nature of reality, or simply strategic deception.
Schema	The conceptual framework or paradigm (of one's mind) contains many constructs or schema, which are subject to periodic revision. See also personal constructs.

Schizophrenia	Major psychiatric disorder characterized by loss of contact with reality, auditory hallucinations and social withdrawal. Such symptoms can be induced in all people through the heavy-handed us of radiofrequency effects – what would differ, of course, is people's individual reactions.
Scientific method	Methodology in which a working hypothesis is tested by experiment. If falsified or refuted a new hypothesis is called-for, otherwise the hypothesis is refined into a theory. The hypothesis that tomorrow the sun will rise can be worked-up into a theory that "every day the sun will rise" (subject, of course, to events in cosmology).
Second-messenger system	A series of enzymatic reaction within a neuron's cytoplasm; reactions which have the effect of releasing energy, sufficient to open sodium ion channels in the neuron's membrane.
Second Sino-Japanese War (1937-45)	War sparked by Japanese imperialism, Japan having invaded Manchuria, a Chinese province, in 1931. The war intensified in 1937, resulting in atrocities on the part of the Japanese; but the Chinese, led by Chiang Kai-Shek, refused to capitulate. Sympathy for the Chinese grew amongst the Western allies, leading to a fateful oil embargo against Japan in 1941. That crippling embargo triggered the Japanese attack on Pearl Harbour; which proved to be Japan's nemesis.
Second World	In the original three world typology which arose after the Second World War, the communist Second World comprised the USSR (and all its allies). Internal differences within the communist sphere of influence led to both the Tito-Stalin split and Sino-Soviet split, fatally weakening the Second World in the process. Such divisions within the Second World contrast with comparative unity within the capitalist First World (Brexit, notwithstanding).
Secretariat (UN)	The United Nations' Secretariat services the principal instruments and bodies of the UN. Its staff answer only to the United Nations, and swear not to seek or receive instructions from any government.
Secretary General (UN)	The chief executive officer of the United Nations, who acts as an ambassador, mediator, negotiator, administrator and envoy.
Secular belief system	Secularism implies that faith and religion should be expelled from public affairs; but that still leaves the question of whether every form of faith should be excluded. Epistemological jurisprudence concedes that not everything derived in an extrasensory manner can or should be proven. Some might learn to trust to that which can't, leading to a secular belief system. See also epistemological jurisprudence.
Security Council (UN)	The United Nations' Security Council is responsible for the maintenance of international peace and security. It both determines and investigates threats to peace, and makes recommendations (such as economic sanctions and/or military action). Its five permanent members (P5) are the five recognized nuclear states, who tested atomic bombs prior to 1968. Additionally, there are ten non-permanent members.
Selective retention	A theory of mass communication which argues that people are most likely to remember information consistent with their existing attitude.
Self (psychology)	The self is that which is derived from ones own person, making an uncompromising defence of the self a defence of personal autonomy and freedom.

Self (surgery)	A surgical term for one's own body tissue, which is recognized by one's immune system, and protected by the same.
Self-actualization	Realizing one's talents and abilities to the full.
Self-concept	A person's perception of their own selves, both in terms of appearance and in terms of disposition, personality and behaviour.
Self-schema	The complete set of memories, ideas, thoughts, feelings and intentions that one holds about oneself. This idealized self may be at odds with spontaneous responses leading to cognitive dissonance.
Semiconservative replication	The name given to the mechanism by which DNA replicates itself. DNA resides in the chromosomes and, provided free nucleotides are available, together with the requisite enzyme (DNA polymerase), energy (ATP), and a moderate temperature, the two helical strands can 'unzip' along their length allowing free nucleotides to attach themselves. Thus, two new double-helixes are formed. Protein synthesis borrows from Semiconservative replication. See RNA.
Semi-isolated system (physics)	A system which is neither open, closed, nor isolated. The human brain attempts to exchange little or no energy with its surroundings, only mass (though it will absorb and emit energy within itself). The human mind is a semi-isolated world, which only humans (as far as we can tell) have the capability to intrude upon.
Sensitive period	The sensitive period for language development is from birth to 6 years of age. Particularly important is the period from 1-3 years when language acquisition is dramatic.
Sex chromosomes	The average human cell has 23 pairs of chromosomes. One of these pairs are the sex chromosomes, which determine the gender of the individual and sexual characteristics associated with that person. The other 22 pairs of chromosomes are common to both sexes.
Sexual selection	An aspect of coevolution in which the whole culture which surrounds sex and sexual attractiveness influences the choice of partners, with that culture (and those choices) then reflected in the gene pool.
Shanghai Cooperation Organization (SCO)	A Sino-Russian power bloc (Russian and Chinese being the official languages), which exists to enhance security between the two states. Its members, together with affiliated nations, are said to comprise half the earth's population. Its importance lies with the fact that China and Russia have permanent seats on the UN Security Council, and many political flashpoints exist within its energy-rich sphere of influence.
Shells (atoms)	The electrons surrounding the nucleus of an atom are restricted to certain orbitals; orbitals which are grouped together into shells and subshells.
Shi'ites	Minority group of Muslims, whose teachings derive from the descendants of the prophet Mohammed. Estimated to represent 10-15% of Muslims, worldwide.
Singapore and Harare Declarations (1971 and 1991)	All members of the Commonwealth of Nations are signatories to the Singapore and Harare Declarations. These documents represent a commitment to good governance, sustainable development, and the promotion of justice, freedom and democracy. However, British arms sales to countries with dubious human rights records makes the charitable work of leading royals look like a public relations exercise.

Single European Act (SEA)	The Single European Act came into force in 1987, with the express intention of completing the internal market between the European Community's member states by 1992 (a market characterized by the free-movement of goods, persons, services and capital).
Singularity	Defined in this book as a point at which all the principal forces in nature are biting. See coupling constant.
Sinn Fein	Sinn Fein, the political wing of the Irish republican Army, refused to accept the partition of Ireland. Consequently both were instrumental in the resurgence of terrorism in Northern Ireland in the 1970s.
Situational attribution	Arises when the cause of a particular form of behaviour is attributed to the person's situation, rather than their temperament.
Six Day War	In 1964 Egypt helped to establish the Palestine Liberation Organization (PLO), leading to an escalation in terrorist activity in the region. In a pre-emptive attack in 1967, Israel destroyed Egypt's air force and then preceded to lay waste to its ground forces. What followed was Israel securing for itself the whole of the original British mandate, in just six exhaustive days.
Social domain theory	According to this theory society comprises three domains: 1. the moral domain; 2. the social-conventional domain; and, 3. the personal domain. Well-balanced societies might appropriate such a model, determining that level one infractions must be corrected through the sparing use of punishment in all places, whereas levels two and three may be accommodated in the interests of multiculturalism, diversity and peace.
Social facilitation	A person can sometimes be rendered more effective when in the company of others, becoming more focused and involved.
Socialist principles	Socialism perceives inequality as socially destructive; and that without equality, self-actualization and the realization of ones potential may be dependant upon inherited wealth, rather than open-competition and merit. Socialist principles stress a reduction in material inequality and access to education and advancement.
Social stratification	Stratification of society based on criteria such as capital, income and professional qualifications.
Sociometry	The systematic or statistical mapping of social networks, so as to build-up a picture of affiliation, association and identification. Such a picture is termed a sociogram.
Sociopathy	Psychiatric condition characterized by violent anti-social behaviour.
Sodium inactivation	The near-instantaneous closure of sodium ion channels in the membrane of a neuron, following the action potential being reached.
Sodium ion channels	Openings in the plasma membrane of a neuron which allow sodium ions (Na^+) to enter the cell triggering depolarization. When this occurs adjacent ion channels open, allowing a nerve impulse to move along the membrane.
Solenoid	A current-carrying coil enclosing an iron core, which can be used as a switch or circuit-breaker. When a current is passed through the coil the iron rod moves, due to the magnetic field created by the current.
Space	Space is an electrically-neutral fabric created from electrically-neutral radiation (radiation whose energy has transduced into the fabric of space in the presence of gravitation).

Space-creation	Space comes into being when the principal forces stop biting, i.e. when the electric, magnetic and gravitational forces fall below a certain magnitude (namely, the coupling constant) and space is the result. As this involves the disintegration of photons, cosmic inflation is a quantum process.
Space-time	Time is a side-effect of space; so that one can legitimately marry the two together, much like electricity and magnetism. However, space gives rise to time, not the other way round.
Special relationship	The relationship which exists between America and Great Britain, due in part to the history of the English-speaking people, but primarily a legacy of their strategic defence of the free world and shared ideals. See also, note 57
Spectroscopy	The systematic study of the properties of light so as to determine information about its source.
Spin	The intrinsic angular momentum of an elementary particle.
Stagflation	Occurs when wages and prices increase during economic slowdown and recession. Quite unlike the Great Depression of the 1930s, which was accompanied by precipitate falls in wages and prices.
Stalinism	Like all one-party communist systems power ultimately comes to rest in the hands of one man, whose personal interpretation of Marxist Leninism becomes de rigueur in that particular state. This was never more so than in the case of Josef Stalin, leader of the Soviet Union from 1922-53.
State	For the purposes of this book, specifically a nation-state or country (as opposed to a state in the USA).
Statute law	Laws which have been enacted by the legislature with the support of the executive branch of government.
Strict liability	Crimes in which only actus reus needs to be proven (in other words, only the act itself).
Sublimation	Occurs when sexual impulses, repressed perhaps through audience effects, surface in non-sexual ways.
Subsidiarity	Principle that decisions should be taken at the lowest appropriate level. In a devolved arrangement the power to make certain decisions is delegated to subordinate regions, but sovereignty remains with that doing the devolving. In a federal arrangement sovereignty is surrendered by various regions (ones delegated specific functions). However, the line between what is devolved and what is federal may become blurred.
Sub-tropical zone	Climatic zone situated between the equatorial zone and the temperate zone (lying approximately between latitudes 10^0 and 25^0 north and south).
Suez Crisis	The Suez Canal, so vital to western interests as a conduit for petroleum, was appropriated by Egypt in 1956, resulting in military action on the part of Britain, France and Israel. Amid fears that the crisis could escalate the United Nations intervened, resulting in the withdrawal of western troops, and the promise of safe passage for western shipping.
Sugar	A sweet-tasting water-soluble chemical substance which comes in a variety of forms. The sugar in DNA is called deoxyribose, which serves as a vital structural component and whose hydrophilic properties helps DNA to adopt and retain its iconic double-helix shape.
Sunni Muslims	Muslims who are adherents of orthodox Islam, whose teachings derive from the prophet Mohammed. Upwards of 85% of Muslims are thought to be Sunnis.

Superego	The human brain's conscience to the ego, affecting and influencing conscious planning and behaviour.
Superimposer	The cognition of one individual can be superimposed upon another provided both their absorption profiles are known. The dominant party (superimposer) is likely to affect the identity and thought processes of the subordinate person.
Superimposition of persons	The projection of one person's neurological processes into another, sufficient to affect innate characteristics and autonomous behaviour. Context is everything, in terms of judging whether this level of imposition is justified or warranted.
Superpowers	In the Cold War era much of the world polarized into pro-American and pro-Soviet factions, reflecting the enormous power and influence that these two geopolitics had acquired. Both the USSR and America were, quite literally, superpowers.
Supranational body	An organization which transcends the politics of individual states, and is more permanent and more structured than intergovernmental associations. For such a body to exist necessitates some transfer of sovereignty to a transnational power.
Supranatural	Supernatural effects, applied by a supranational organization. Such effects are experienced in an extrasensory manner.
Supranatural selection	Natural selection occurs because some individuals within a species are less affected by environmental resistance (due to advantageous mutations); those individuals becoming, in time, a new species type. Global politics has similarly evolved, with an invisible hand favouring the Allies over the Axis Powers, the Western Allies over the Soviets, and due process over the alternatives.
Systems theory	The earth and its inhabitants comprise so many discrete but nonetheless interconnected systems. These systems form branches within science, which reductionism breaks down into smaller, ultimately quantized pieces. Systems theory attempts to make sense of these relationships, in order to make predictions about the dynamical whole.
T	
Taliban	Islamic revolutionary movement which came to power in Afghanistan in the decade following the withdrawal of Russian forces in 1989. Under the Taliban strict sharia law was imposed, with men obliged to grow beards and women forced to wear the burqa. An American-led full-scale invasion of Afghanistan in October 2001 displaced the Taliban.
Telekinesis	The manipulation of objects using only the power of one's mind.
Telepathically-induced effects	The range of effects achievable by means of radiofrequency technologies, i.e. remote manipulation, mental telepathy, and the superimposition of persons. Additionally, one might add psychokinesis and telekinesis. Volume two establishes a purpose-built clinical model to explore these effects – that clinical model comprising two divergent cycles, whose radiological, electrochemical and hormonal properties combine, with the aid of language, to produce four discernible neurotypes.
Telepathy	See mental telepathy.

Temperate zone	Climatic zone situated between the sub-tropical zone and the boreal zone (lying between latitudes 25^0 and 50^0 north). In the southern hemisphere there is no boreal zone, it being almost exclusively ocean, so the southern temperate zone terminates with open sea. Note: this zone contains many arid and semi-arid areas.
Tentative answer	Working hypotheses that have not been falsified or rebutted, in spite of rigorous tests and observation, produce tentative answers.
Terrorist attacks, 9/11 (2001)	Attacks masterminded by Osama bin Laden, and executed by members of the al-Qaeda terrorist organization. Nineteen Arab terrorists hijacked four planes, three of which smashed into their intended targets – the fourth of which crashed short of its objective. Around 3,000 Americans lost their lives in these attacks; attacks which completely destroyed New York's World Trade Centre site.
Thanatic	See Thanatos.
Thanatos	In Freudian psychology thanatos represents a destructive potential, which – like molecules and the complex magnetic forces they give rise to – resides within all of us; and which may build and be amplified in groups (amplified due to people having a similar orientation).
Theism	The belief that there is a supreme deity or God.
Third World	With fascism confined to history (following the conclusion of the Second World War) the ensuing Cold War became a competition between pro-American and pro-Soviet factions. Many underdeveloped nations in the southern hemisphere remained neutral or non-aligned, becoming known as the Third World.
Time	Ontology questions what actually exists. This book argues that all that exists requires space, including time itself. So time began with the creation of space, and will end only when space ceases to exist.
Titoism	Josip Broz Tito's communist Yugoslavia created a near counter-revolutionary model, free of the overt imperialism of the Soviet Union, and not adverse to pragmatic dealings with the West.
Tito-Stalin split (1948)	Communist Yugoslavia, under Josip Broz Tito, became distanced from the USSR, due to irreconcilable differences between Tito and Josef Stalin, the General Secretary of the Communist Party of the Soviet Union. Those differences and the rift which followed served to make the world a safer place, as it isolated the Balkans region.
Tort (Law)	A wrongful, unwarranted and illegal act for which one might be fairly compensated. Torture, for example, is both a crime and a tort.
TPP	The Trans-Pacific Partnership (TPP) currently comprises Brunei, Chile, New Zealand and Singapore; but is looking to incorporate the whole of North America, Australia and Peru.
Transduced	The conversion of energy, from one form into another.
Transistors	In 1947 transistors replaced valves in most electronic devices, serving much the same function, whilst being smaller and more reliable. These semiconductors were capable of extracting, rectifying and amplifying radio signals more effectively than previous technology.

Transmittance	Radiation passing directly through an object, measured as a ratio of transmitted radiation to incident radiation.
Transmitter	The apparatus needed to generate a radiofrequency wave; which is modulated, so that meaningful information is conveyed to those receiving the same.
Transverse waves	A wave which affects the medium through which it travels, in a way which is at right angles to the direction of travel.
Treaty of Athens (2004)	Treaty which provided for the accession of ten former Warsaw Pact countries to the European Union (all former republics of the Soviet Union). Bringing the number of countries in the EU to twenty-five.
Trias politica	The political system based upon the separation of power between three distinct bodies; namely, the executive, legislature, and judiciary. The trias politica arrangement, in combination with radiofrequency effects, accounts for the First World growing in strength and stability (the Second and Third Worlds' having suffered splits and stagnation). That said, Brexit reveals a stubborn determination to weaken the First World, thereby jeopardizing NATO as the post-peak future approaches.
Trieste Accords	Treaty brokered by the Americans and British between Yugoslavia and Italy in 1954, which provided for the city of Trieste remaining in Italian hands, whilst the surrounding countryside passed to Yugoslavia.
Trinity Test	The world's first successful testing of a nuclear bomb, carried out by America's top-secret Manhattan Project on July 16, 1945.
Trusteeship Council (UN)	The United Nations' Trusteeship Council assisted trust territories (i.e. former colonies and overseas possessions) to achieve self-governance and independence in the postwar era.
Turner Prize	Prestigious art award, presented each year by the Tate to a British artist who has made an outstanding contribution to contemporary art in the preceding 12 months. As a mechanism for provoking debate about art, it has no equal.
Two-step flow theory	A theory of mass communication which argues that intermediaries heavily mediate the message coming from primary sources, thereby affecting our overall understanding.
U	
UK	See United Kingdom.
UK Supreme Court	United Kingdom's highest court, based in London.
Uncodified	Often used to describe the constitution of a country, whose constitution isn't formally written down in a single comprehensive document, but rather comprises certain laws, conventions and precedents.
Unconscious competency	The ability to perform a task subconsciously, perhaps whilst doing other things, due to the task being imprinted on the nervous system (in the manner of an efference copy).
Union of Soviet Socialist Republics (USSR)	Former communist federation arising after the Russian Revolution (1917), and comprising various republics, e.g. Russia, Ukraine, Siberia, etc. The USSR greatly extended its influence after World War Two, due to the Warsaw Pact (1955), which created a corridor of Soviet-dominated countries between the USSR and Western Europe.

United Kingdom (UK)	England (which emerged from the heptarchy of the Dark Ages) formerly annexed Wales through the Act of Union of 1536. England and Wales were then joined by Scotland, via the Act of Union of 1707, creating Great Britain. Great Britain subsequently passed the Act of Union of 1800, thus incorporating the whole of Ireland, whereupon these islands became a United Kingdom. The partition of Ireland, which arose due to the Anglo-Irish war of 1919-21, created a United kingdom comprising the British mainland and Northern Ireland. Devolution, beginning in the late 1990s, then provided for assemblies and parliaments in Northern Ireland, Scotland and Wales. In Constructive Interference: Developing the brain's telepathic potential (aSys Publishing, 2019) we see how the United Kingdom can't succeed post-Brexit if the body politic remains largely sub-aequipine – that is to say, the UK (like the EU and USA) must somehow address the corrosive impact of Homo sapiens.
United Nations (UN)	Formed in 1945, the United Nations is tasked with preserving global peace and security, and with facilitating cooperation between member states. Its six principal bodies are: 1. General assembly; 2. Security Council; 3. Secretariat; 4. International Court of Justice; 5. Economic and Social Council; and 6. Trusteeship Council.
Universal suffrage	The right to vote in general or presidential elections – when extended to every person of full age, irrespective of gender – is termed universal suffrage.
UN Security Council	A principal instrument of the United Nations, the Security Council comprises five permanent members and ten non-permanent members. Together they are responsible for the maintenance of international peace and security. Additionally, they recommend the appointment of the UN Secretary General.
US Constitution	The articles binding the thirteen American colonies were formally replaced by the US Constitution in 1789. It contains seven articles and twenty-six amendments.
USSR	See Union of Soviet Socialist Republics.
US Supreme Court	America's highest court, based in Washington DC.
V	
Valence electrons	Those electrons located in the outermost shell of an atom, which are directly involved in chemical bonding.
Valence shell	The shell containing the outermost electrons. These valence electrons are decisive in determining an atoms tendency to bond with other atoms.
Valves	From 1904 valves became available, which exploited newly-discovered cathode rays (i.e. electrons), which passed in one-direction only through glass tubes. By this means the amplification of radio signals became possible.
Vicarious liability	Criminal responsibility often rests with the person who committed an infraction, but in the case of vicarious liability that culpability extends to others, e.g. employers, bureaucrats and even the state. Thus criminal responsibility for the crime of genocide extends well beyond the immediate perpetrators.
Viet Cong	Communist Vietnamese, with pro-Soviet sympathies.
Viet Minh	Vietnamese nationalists, driven by a desire for self-determination, who accepted Soviet assistance as a means of achieving self-governance.

Vietnam War (1964-75)	American, Soviet and UN intervention in the internal affairs of Vietnam led to the country's partition in 1954. The controversial Gulf of Tonkin incident, which amounted to alleged aggression by the Soviet-backed north, sparked a growing conflict between North Vietnam and the American-supported south. Despite a massive commitment by American armed forces the south lost, resulting in the domination of the entire country by Soviet-assisted nationalists.
Visible light	That portion of the electromagnetic spectrum which is visible to humans. Convergence (as regards the evolution of colour vision across numerous species) suggests that beyond the visible electromagnetic radiation is either too hazardous, too intermittent, or completely absent.
Voting Rights Act (1965)	Landmark civil rights legislation in the US, which had the effect of enforcing the 15th Amendment to the American Constitution – which had stated that a US citizen's right to vote should not be denied on account of race, colour or previous condition of servitude.
W	
Wall Street Crash (1929)	Following the American Civil War (1861-65) there was a wholesale rationalization of the US banking system, and the provision of a uniform currency. Further reforms, including the introduction of the Federal Reserve System (1914), added to America's financial importance globally. The collapse of New York's Stock Exchange, better known as the Wall Street Crash, in 1929, confirmed that global importance, as the flow of capital out of America ceased (with dire repercussions for economies worldwide).
Wannsee Conference (1942)	By the summer of 1941 the Nazis had moved towards the mass-extermination of European Jews. The Wannsee Conference, chaired by Reinhard Heydrich, in 1942, sought to make this killing machine ever more efficient through the establishment of death camps, mass transportation and the pitiless use of gas.
Warsaw Pact Countries	The East European Mutual Assistance Treaty was signed in Warsaw in 1955; its signatories included East Germany, Poland, Czechoslovakia, Hungary, Romania, Bulgaria, and the USSR. This unified military bloc arose in the aftermath of NATO's creation.
Wavelength	The distance between two successive crests of a wave, along its line of travel (expressed in metres).
Wavenumber	A 10 metre wavelength (1000 cm) would have a wavenumber of 0.001 cm^{-1} (i.e. the number 1 divided by the wavelength, expressed in cm). In other words, the wavenumber is the reciprocal of the wavelength.
Wave-particle duality	That light propagates in the manner of transverse waves, but interacts with matter in the manner of particles.
Well-balanced	Conforming to the rule which says that one should favour rewarding; using punishment sparingly, and only to define the limits.
West, The	Abbreviation of Western Allies.
Western Allies	Those fighting on the Allied side in World War Two, who went on to form the North Atlantic Treaty Organization (NATO).
Western Polyarchies	NATO affiliated nations with representative democracies, capitalist economies and a commitment to liberal individualism.

White Paper of 1922	White Paper which codified Britain's commitment to the establishment of a Jewish homeland in the mandate of Palestine.
Wikileaks	In 2011 the non-profit organization Wikileaks began publishing classified material relating to the alleged maltreatment of Guantanamo Bay prisoners (i.e. terrorist suspects captured in places such as Afghanistan). Theories of mass-communication suggest that people tend to remember information consistent with their existing attitude, so fundamentally changing attitudes can be difficult, even with this level of awareness.
World Trade Centre	Complex formerly located in lower Manhattan, New York, and dominated by its two imposing 110-storey twin towers. The whole complex was destroyed in the 9/11 terrorist attacks, in 2001.
X	
X-ray crystallography	A technique in which X-rays are directed at a solid sample, possessing a regular or crystalline structure – the way in which those X-rays are scattered by the atoms tells us something about its structure.
Y	
Yellow star	Distinguishing mark, fashioned from yellow cloth, cut into a star of David, which European Jews were increasingly forced to wear following the Nazi occupation of Europe. This was done with the express intention of marking them out for discrimination, humiliation and deportation. See also Anti-Semitism.
Yom Kippur War	In 1973, newly equipped with powerful Soviet weapons, the Egyptian armed forces dared to launch a full-scale assault on lands procured by the Israelis in the Six Day war. Combined pressure from the Americans, Soviets and United Nations resulted in no significant territorial changes.
Z	
Zionists	Those who argued for the establishment of a permanent Jewish homeland in British controlled Palestine.

274

NOTES

1   As regards natural forces, Cyril Aydon writes: "Science is humanity's attempt to explain them. Technology is humanity's attempt to exploit them". *The Science Book* (Magpie, 2010), p.46.

2   Frederic Raphael, *Popper* (Phoenix, 1998), p.5, states "Any idea that cannot conceivably be refuted is not scientific." Précis of Karl Popper's life and work.

3   Desmond M Burns and Simon GG MacDonald, *Physics for biology and pre-medical students* (Addison-Wesley Publishing Company, 1970), '20.2 Interference' p.302, explains that only by assuming a wave motion can interference be explained.

4   As furbish is a rarely used lexeme, this book generalizes its meaning, making it synonymous with *contemporize*; with furnish meaning to *equip* or *supply*.

5   The significance of Neils Bohr's *On the Constitution of Atoms and Molecules* is vividly conveyed in Piers Bizony's book, *Atom* (Icon Books, 2008), p.36. Also, for an illustration of the Bohr model of the atom, see Adam Hart-Davis (Editor-in-Chief), Science (Dorling Kindersley Limited, 2010), p.287.

6   Whilst the orthodoxy maintains that like charges repel, Constructive Interference: Developing the brain's telepathic potential (sAys publishing, 2019) critically evaluates that claim.

7   Why didn't the big bang create much heavier elements? For an alternative astrophysical explanation, see Mark Fox, *Constructive Interference: Developing the brain's telepathic potential* (aSys Publishing, 2019), p7.

8   "Feynman wanted to cut through the infinities..." is a verbatim quote from Piers Bizony, *Atom* (Icon Books, 2008), p.145. In this book, Bizony asks: "Are fields real?", p.141 – *Destructive Interference* suggests they are, e.g. a magnet's fields align and strengthen sufficient to pick-up a paperclip. In other words, if fields aren't real, what is aligning and strengthening?

9   Figure 3, an example of a Feynman diagram, appears in Professor Peter M B Walker, CBE, FRSE (General Editor), Dictionary of Science and Technology (Larousse plc, 1995), p.414. Clearly, whenever one is faced with two particles interacting across *space* and *time* the question arises as to whether those particles communicate solely by means of other particles. *Destructive Interference* perceives *space* and *time* as arising due to particles failing to interact strongly enough, and that fields are an integral part of that dynamic.

10   HJP Keighley, FR McKim, A Clark and MJ Harrison, *Mastering Physics* (The MacMillan Press Ltd, 1982), p134, explains the difference between constructive and destructive interference.

11   David Crystal (Editor), *The New Penguin Encyclopedia* (Penguin Books Ltd, 2002), 'interference' p.772, explains what steps must be taken to observe interference patterns under controlled conditions.

12   The role of water and aqueous solutions in the human body is discussed in J. Simpkins and J.I. Williams, *Advanced Human Biology* (Unwin Hyman, 1987), p.1-8.

13   The comment "...the principal aim of chemical bond formation could be considered to be the generation of full valence shells" comes from Jonathan Crowe, Tony Bradshaw and Paul Monk's Chemistry for the Biosciences (Oxford University Press, 2006), p.37. A first-rate introduction to the chemistry underlying the life sciences.

14   Whereas this book focuses on the intrigues surrounding the establishment of today's scientific orthodoxies, the second book – *Constructive Interference: Developing the brain's telepathic potential* (aSys publishing, 2019) – supplants those incommensurable models and replaces them with a whole new paradigm (i.e. this book's *first edition* was weakened by those orthodox arguments, thereby necessitating this timely *second edition*).

15   Virginia Woolf was treated for psychosis and depression, eventually committing suicide in 1941. Kay Redfield Jamison, *Touched with Fire* (Free Press, 1994), p.225, quotes her as follows, "I am a porous vessel afloat on sensation…a sensitive plate exposed to invisible rays…taking the breath of these voices in my sails…".

16   J.O. Urmson (Ed) and Jonathan Rée (Ed), *The Concise Encyclopedia of Western Philosophy and Philosophers* (Unwin Hyman, 1991), p.172, defines libertarianism as a rejection of determinism, with the criticism that if human behaviour *isn't* determined by inherited character or conditioning, what alternative is there. The writer has appropriated this term, as meaning a rejection of externally-imposed controls and intrusion, initiated by others.

17   'Affect' is defined as *"to act upon or influence, especially in an adverse manner"*. Ergo, 'affectation' implies superimpositions with undesirable ramifications.

18   A summary of the main structural differences so far observed between men's and women's brains can be found in Rita Carter's *Mapping the Mind* (Weidenfeld & Nicolson, 1998), p.71. Additionally, Anthony Smith's *The Human Body* (BBC Books, 1998), p.123, states "Men have, on average, larger brains than women, but women's brains occupy a greater proportion of their bodies."

19   An introduction to perception is given in Michael W Eysenck, *Principles of Cognitive Psychology* (Lawrence Erlbaum Associates, 1994), p.40. Additionally, *The Britannica Guide to the Brain* (Robinson, 2008), p.63, explains the difference between sensing and perceiving.

20   The second book, *Constructive Interference: Developing the brain's telepathic potential* (aSys Publishing, 2019), introduces a hierarchy of neurotypes, thus aiding that concentration of knowledge, power and perceptiveness.

21   For definition of legitimate power, expert power, informational power, etc, see Nicky Hayes, *Principles of Social Psychology* (Lawrence Erlbaum Associates, 1995), p.150-160.

22   "…exaggeration of deeper imbalances of power" comment, prompted by Natasha Walter's book, *Living Dolls – The Return of Sexism* (Virago Press, 2010), p.8.

23   Denial, as a common denominator across all societies is the subject of Stanley Cohen's book, *States of Denial – knowing about atrocities and suffering* (Polity Press, 2008). Useful primer, when thinking about informational power.

24   The origins of civilization, and the importance of trade and commerce in particular, are discussed in the BBC *Horizon* programme, *The Lost Pyramids of Caral*, 2002.

25   "Rousseau argued that only through the direct and continuous participation of all citizens in political life can the state be bound to a common good, or what he called the 'general will'…" quote taken from Andrew Heywood, *Politics* (MacMillan Press, 1997), p.8.

26   The second book classifies individuals on the basis of their cognitive, nervous and endocrinal responses, with NEURON being 'populated' by **balanced** persons. As for the mainstream, the *international scientific community* would consist largely of **sound** minds, with **logical** and **biased** mindsets actively distanced from NEURON. See *Constructive Interference: Developing the brain's telepathic potential* (aSys Publishing, 2019), p136.

27   The writer submitted a 'blog pitch' to the Feminist Studies Association in 2020, entitled *Child, Early & Forced Marriage: A Crime of Omission?* Central to that pitch was the question of whether harmful practices, in respect of women and girls, could be eliminated without recourse to international criminal law, *ad hoc* tribunals and omission liability. That '**blog pitch' is reproduced in full at the end of this book.**

28   The 'Three Worlds' typology, or classification, is outlined in Andrew Heywood, *Politics* (MacMillan Press, 1997), p.27. Human intelligence is the ability to pro-actively learn and suitably adapt – accordingly, First World countries with Left-wing sympathies are manifestly more intelligent than either the Second or Third Worlds (and, indeed, Right-leaning First World countries) due to their conscientious appraisal of extrinsic concerns.

29   *The House of Saud* (Alegria/Starling co-production for the BBC, 2004). Important documentary as regards the history of Saudi Arabia and the emergence of al-Qaeda. Written, directed and narrated, by Jihan El Tahri. *Destructive Interference* draws heavily on those arguments expounded in this superlative documentary (more specifically, within the section entitled 'Neuro-cognitive power and the rise of al-Qaeda').

30   Leslie Robertson, the lead Structural Engineer on the Twin Towers project, is quoted verbatim from a TV documentary, produced and directed by Garfield Kennedy & Larry Klein, *Why the Towers Fell* (BBC / WGBH Boston co-production), 2002.

31   "On September 11th, the United States, so proud and so free..." quote taken from *The Day that Shook the World – Understanding September 11th* (BBC Worldwide Ltd, 2001), p.145, "Israelis and Palestinians in the Aftershock", chapter written by Orla Guerin (edited by Jenny Baxter and Malcolm Downing).

32   The axiom or rule, referred to in the section 'Impart answer', comes originally from childcare, and was encountered whilst browsing the shelves of Bookcase, Carlisle. The writer has appropriated this model of good practice in childcare as being the exemplification of balanced-mindedness, thereby adding to its wider utility and importance. This model then forms the basis of those arguments encountered in the section 'Anthropobionomics' (Ch.10), i.e. that modern humans sought to maximize the rewards associated with increasingly higher latitudes, whilst reducing, as far as possible, the punishing effects of longer colder winters and increasingly shorter cooler summers.

33   Professor Jared Diamond, *Guns, Germs and Steel* (Lion Television Production for National Geographic and Channel 4 Television, 2005) states that: "...at roughly the same line of latitude any two points on the globe automatically share the same length of day, *and* similar climate and vegetation". The writer of *Destructive Interference* concludes that if modern humans reached Australia before Europe, as suggested by Dr Alice Roberts, *The Incredible Human Journey* (BBC Worldwide Ltd, 2009), then one might propose that early modern humans sought to avoid having to innovate and adapt, primarily by remaining within much the same latitude.

34 The sudden escalation in anti-Jewish killings, in the summer of 1941, is analyzed in Laurence Ree's seminal work, *The Nazis – a warning from history* (BBC Worldwide Ltd, 1997), p.194. This book also mentions Hans and Sophie Scholl, students at the University of Munich, who courageously produced leaflets during the war calling for German youth to rise up, in order to build a new spiritual Europe. Both Scholls were later denounced, tortured and murdered by agents of the Third Reich.

35 An overview of the events surrounding the deportation and murder of Hungary's Jewish population during the Holocaust can be found in Professor David Cesarani's introduction to *The Last Days* (Seven Dials, 2000), p.15-51.

36 Professor Tony Judt's comments, regarding the "scale of punishment meted out to the citizens of the USSR..." are taken from his book, *Postwar* (William Heinmann, 2005), p.191. Judt says : "In 1952, at the height of the second Stalinist terror, 1.7 million prisoners were held in Soviet labor camps, a further 800,000 in labor colonies, and 2,753,000 in 'special settlements'. The 'normal' Gulag sentence was 25 years, typically followed (in the case of survivors) by exile to Siberia or Soviet Central Asia".

37 *The New International Webster's Pocket Grammar, Speech and Style Dictionary of the English Language* (Trident Press International, 2001), p.72, states: "It has been said that there is no such thing as good writing, only re-writing. All really great authors polish their works many times".

38 George Orwell's novel, *Nineteen Eighty-Four*, was first published by Martin Secker & Warburg in 1949.

39 Lee Harvey Oswald fired three rifle bullets in the course of President Kennedy's assassination; however, the first of these missed the President completely, doing no more than drawing Governor Connally's attention. The second and third bullets, however, proved fatal.

40 Professor Tony Judt, *Postwar – a history of Europe since 1945* (William Heinmann, 2005), p.736, "In 1980 the sum of all international bank lending was $324 billion a year; by 1991 that figure had grown to $7.5 trillion – a 2,000 percent increase in just over a decade."

41 Professor Niall Ferguson, *Colossus* (Penguin Books, 2005), p.273, states "...Thatcher reforms are the reason the United Kingdom is one of the elite of the developed economies that do not have major holes in their generational accounts." [*2021 Covid-19 addendum: "they do now!"*]

42 BBC's Newsnight (broadcast 03.09.2004) reported on the Beslan School Hostage Crisis, showing Regional Federal Security Chief, Valery Andreyev, stating "...we never planned any forced action". In the sudden melee which surrounded the end of the siege many Russian Special Forces went in without the benefit of distinguishing armbands.

43 The Arabs were nominally allied to the Axis Powers according to several sources, most notably Robert Fisk, *The Great War for Civilisation* (Fourth Estate, 2005), p.440-448, in which he quotes Haj Amin as saying on Nazi radio, "The Germans know how to get rid of the Jews...they have definitely solved the Jewish problem".

44 Panorama, *Killers* (BBC, 1994/2004). Reporter, Fergal Keane's timely examination of the aftermath of the Rwandan genocide includes a former public official saying "You could argue many reasons as to why they did it – but they are all tied to power. You know how power is sweet. That is the root of the genocide."

45   Professor Niall Ferguson, *Colossus* (Penguin Books, 2005), p.97, recites former US Secretary of Defense James Schlesinger: "...rather than simply counter your opponent's thrusts, it is necessary to go for the heart of the opponent's power..."

46   One's heteronnubial co-author – whose intelligence, acumen and judgement exceeds that of the writer – contends that by not removing Saddam Hussain, in the First Gulf War, the USA made the 9/11 attacks more likely.

47   The Victorians were better educated than their forebears, making them acutely conscious of the violence inflicted by the Anglican and Roman Catholic churches – hence the proliferation of Celtic crosses which suddenly appear in English churchyards in the late 19th century (the Celtic church being, to all intent and purposes, a better exemplar of virtue, meekness and restraint).

48   Dr Shompa Lahiri, in the chapter entitled "The Aftermath of Empire", Atlas of British & Irish History (Penguin, 2001), p.261, makes reference to the 1948 Nationality Act and subsequent legislation.

49   It could be argued that England always was, and always will be, a 'European project', its roads, towns, cities, language, law, architecture, etc, all being European hand-me-downs, gifted to us by the Romans, French, Saxons, Angles, Jutes, Danish, Angevins, etc (even the monarchy itself is a European import, incorporating hand-picked immigrants).

50   Figure 22, America at the time of the Civil War, is an original image, produced by the writer, using information from a contemporaneous source, i.e. it utilizes a map produced by a soldier working for the US Federal Government, entitled: *The Civil War 1861-1865*.

51   Constitutionally, one might question the wisdom of swearing allegiance to a monarch who presides over 15 *independent sovereign states* (excluding Barbados, which is set to become a republic), in-so-far as it creates a *conflict of interest*, e.g. where their elected governments differ, where the monarch gains financially or where classified information is involved. In principle, it makes more sense to have the '*russet-coated*' swear allegiance to parliament, which is the mainstream's ultimate arbiter on matters of state, including the laws themselves – rather than to the head of a foreign realm, such as Belize.

52   For concise definitions of specific legal terms, such as *habeas corpus* and *actus reus*, etc, the reader may wish to consult Graham Gooch and Michael Williams, *Oxford Dictionary of Law Enforcement* (Oxford University Press, 2007). However, *Destructive Interference* expands upon these terms and in some instances broadens their meaning.

53   An unsound phenomenon triggers uncharacteristic behaviours – behaviours which then compete for people's attention, often to the detriment of life-preserving activities affecting far greater numbers of people.

54   One of contemporary society's most pernicious concerns is *small-mindedness*, whereby comparative trivialities deflect people from resolving far graver challenges. According to that argument, an *agent provocateur* need only deflect people trivially in order to inflict catastrophic harm.

55   Douglas Hurd, *The Search for Peace* (Little, Brown and Company, 1997), p.58, "Suppose that Britain and France had gone to war with Germany when Hitler occupied the Rhineland. It is clear now...that the Allies would have won".

56   The section 'Peak religion and its secular aftermath' begs the question, what is agnosticism? In answer, the second book views agnosticism as an open-mindedness as regards the sum impact of all *electromotive* and *nucleomotive* forces.

57   The United Kingdom's single most important strategic ally is NATO – after all, the UK 'stood alone' after being attacked by Nazi Germany! Ergo, the 'NATO commitment' is strategically more important than a nuclear deterrent, and far more cost effective.

58   *The Duel*, a documentary written and directed by Simon Berthon (3BM television production for Channel 4), quotes Sir Winston Churchill as anticipating the new world order with his comment "There would be a United States of Europe, and this Island would be the link connecting this Federation with the New World and able to hold the balance between the two..." *Destructive Interference* takes the view that these entities should serve the free world, irrespective of whether that necessitates merging or remaining discrete geopolitics.

59   Figure 31, NATO members and global partners, derives from information appearing on the NATO website in 2013 (http://www.nato.int/cps/en/ natolive/index.htm). This website lists both NATO members and global partners – including those countries allied via the Euro-Atlantic Partnership Council, Mediterranean Dialogue, Istanbul Cooperation Initiative, Partners Across the Globe, and through EU membership.

60   There are hazards associated with borrowing terms in common usage, 'coupling constant' being one of them – that said, there's symmetry in all the examples given, placing constants at the heart of all such changes. In the next book, we examine what happens when a field encounters an unidentified symmetry, and the changes which then follow, e.g. the field responds *either* (a) galactically, *or is* seen to respond (b) intergalactically (with that very symmetry signalling the conservation of mass and energy, sufficient to bring the 'big bang' hypothesis into question).

61   Both Coulomb's Solution and the Dark Cycle are the writer's own invention. Implicit in *Destructive Interference*, however, is the suggestion that science may be withholding such truths – for example, due to political and military intrigue or in deference to the scientific method and double-blind controls.

CHILD, EARLY & FORCED MARRIAGE: *A CRIME OF OMISSION?*

- *Turning-up the heat on those failing to eliminate harmful practices in respect of women and girls*

As there is *"no specification on the minimum age of marriage in international law"*[1], Her Majesty the Queen, as **Head of the Commonwealth**, appears reluctant to insist on a commonly agreed minimum age of marriage across all Commonwealth Nations. The question, of course, is whether the Commonwealth institutes that change pro-actively, or – as seems likely – they are compelled to do so by virtue of *international law*. Indeed, whilst there doesn't appear to be a precise specification on the minimum age of marriage in international law, the United Nations' **Convention on the Rights of the Child** (1989) expressly states that *"a child means every human being below the age of eighteen years"*[2], with the **Convention on the Elimination of All Forms of Discrimination against Women** (1979) stating that *"all necessary action, including legislation, shall be taken to specify a minimum age for marriage"*.[3] In other words, one might have expected the Queen to have done more to protect the interests of children, in deference to those directives, rather than simply side-stepping the accusation of having committed a *crime of omission*[4], ostensibly through disingenuous phraseology.

Building upon the principle that *"command responsibility is the only true form of omission liability in international criminal law"*[5], one proposes that the legal definition of *command responsibility* be extended, sufficient to criminalize heads of state, government officials and the leaders of intergovernmental organizations.[6] For example, for the last two years Boris Johnson has been the Commonwealth's Chair-in-Office[7], making his merging of the **Department for International Development** (DfID) and **Foreign & Commonwealth Office** (FCO) controversial, due to fears that development aid will be used for *political* rather than *humanitarian* purposes,[8] thereby aggravating the social and economic factors driving child, early and forced marriage (with a no-deal Brexit further weakening the **Foreign, Commonwealth and Development Office's** (FCDO) commitments in those humanitarian areas not directly impacting upon the UK). That is to say, in spite of Boris Johnson's prestigious role in the day-to-day running of the Commonwealth, there's every reason to believe he'll ignore those factors driving child, early and forced marriage.

If – as the growing raft of agreements in respect of child, early and forced marriage suggests – there is a change in *international law*, it will be thanks to the **United Nations** (whereupon Her Majesty the Queen, as Head of the Commonwealth, will find themselves under pressure to act, or face the accusation of having committed a *crime of omission*, they having notionally frustrated the *sustainable development goal* of eliminating child, early and forced marriage by 2030).[9] In point of fact, the charitable organization Girls Not Brides (which is *"a global partnership of more than 1400 civil society organisations committed to ending child marriage and enabling girls to fulfil their potential"*) states that 193 countries have agreed to end child marriage by 2030.[10] However, it is by no means clear whether that aspiration can be achieved without recourse to *international criminal law, ad hoc tribunals* and *omission liability*. Undoubtedly, given that *"all necessary action"* hasn't been taken at the highest levels, there are many reasons to believe that child, early and forced marriage can *only* be eliminated with reference to the same.

Unsurprisingly, the **Commonwealth Secretariat** can provide numerous social and economic explanations for why child, early and forced marriages occur – notwithstanding that focusing on the social and economic reasons for homicide, for example, in the absence of unambiguous legal safeguards, is tantamount to murder. As things stand, the situation has the appearance of the Crown abnegating its criminal responsibilities in – of all things – *international law*. That deficiency being further aggravated by the **Right-wing media**, whose practice of scapegoating celebrities often serves to obscure far more pernicious concerns.[11] In other words, those high-profile 'witch hunts' – so vaunted by today's media – often leave the general public obscenely acquiescent, if not actively complicit, in a litany of harmful practices, many affecting women and girls (with *criminal omission*, in respect of the United Nations' *sustainable development goals*, appearing more likely as a result).

Today, we find that several Commonwealth Nations (such as Zambia, Sierra leone, Namibia, Cameroon and Rwanda) have *"no minimum legal age of marriage"* – placing them on a par with the worst of Arab Gulf States (namely, Yemen, Oman and Saudi Arabia).[12] And yet, the very fact that Article 1 of the United Nations' **Convention on the Rights of the Child** (1989) expressly states that *"a child means every human being below the age of eighteen years unless under the law applicable to*

282

the child, majority is attained earlier[13], strongly suggests that an inexcusable incongruity has been allowed to perpetuate itself. With that oversight appearing even more incomprehensible when one considers that the **Convention on the Elimination of All Forms of Discrimination against Women** (1979) specifically states that "*The betrothal and the marriage of a child shall have no legal effect, and all necessary action, including legislation, shall be taken to specify a minimum age for marriage and to make the registration of marriages in an official registry compulsory*".[14] That is to say, given those longstanding obligations, hasn't the **English Crown** a case to answer in respect of *omission liability*?

Intelligence, one argues, is the ability to "*pro-actively learn and suitably adapt*" – such that the unfortunate female victims of child, early and forced marriage are customarily denied adequate opportunity to intelligently express themselves – and, it must be said, for the most unintelligent of reasons, i.e. an unsound patriarchy has failed to learn from the growing list of statutes; is constitutionally incapable of adapting to the same; and harbours far too many individuals, of both sexes, who are guilty by omission. Evidently, no one is above the law – not even Her Majesty the Queen – a conclusion which should prove sobering to a host of Commonwealth officeholders, high-ranking bureaucrats and government officials (and, indeed, anyone else who feels that human rights abuses can, in some way, be politely explained away). And so, whilst an '*International Criminal Tribunal in respect of crimes against the child within the Commonwealth of Nations*' appears remote, it's not, and never will be, altogether impossible.

SOURCES

1. www.commonwealthroundtable.co.uk/commonwealth/child-not-child-questions-commonwealth-wide-overview

2. https://www.ohchr.org/en/professionalinterest/pages/crc.aspx

3. https://www.un.org/womenwatch/daw/cedaw/cedaw.htm

4. https://www.oxfordreference.com/view/10.1093/oi/authority.20110803100249628 (defines 'crime of omission')

5. Gideon Boas, 'Omission Liability in International Criminal Law - A Case for Reform', in Shane Darcy and Joseph Powderly, Judicial Creativity at the International Criminal Tribunals (Oxford University Press, 2011), 204-226

6. en.wikipedia.org/wiki/Command_responsibility "*command responsibility is an omission mode of individual criminal liability: the superior is responsible for crimes committed by his subordinates and for failing to prevent or punish (as opposed to crimes he ordered)*".

7. https://thecommonwealth.org/about-us/how-we-are-run ("*The United Kingdom's Prime Minister Boris Johnson is the current Commonwealth Chair-in-Office*", 11.09.20)

8. https://www.theguardian.com/global-development/2020/jun/16/political-vandalism-dfid-and-foreign-office-merger-met-with-anger-by-uk-charities (Stephanie Draper, speaking for international development NGOs, said that keeping an independent DfID is the best way to ensure aid is spent on "*helping those most in need, delivers impact for the British taxpayer and remains untied to our political interests*")

9. https://www.girlsnotbrides.org/themes/sustainable-development-goals-sdgs/ ("*target 5.3 aims to eliminate all harmful practices, such as child, early and forced marriage and female genital mutilations by 2030*")

10. https://www.girlsnotbrides.org/themes/sustainable-development-goals-sdgs/ ("*ending child marriage will help us achieve at least eight of the SDGs, and 193 countries have agreed to end child marriage by 2030*")

11. An email dated 24.05.20, entitled "*Unmissable Media-led Smokescreens*", was forwarded by the writer to Tracy Brabin MP (raising concerns regarding "*proliferating media circuses*")

12. https://www.girlsnotbrides.org/where-does-it-happen/atlas/ (02.09.20)

13. https://www.unfpa.org/child-marriage-frequently-asked-questions "Is child marriage legal?", and https://www.ohchr.org/en/professionalinterest/pages/crc.aspx

14. https://www.un.org/womenwatch/daw/cedaw/cedaw.htm

Article written and researched by Mark Fox BSc (Hons) First Class, 2020

www.ingramcontent.com/pod-product-compliance
Lightning Source LLC
Chambersburg PA
CBHW071332210326
41597CB00015B/1427